# THE BREAKDOWN OF DEMOCRATIC PARTY ORGANIZATION 1940-1980

# The Breakdown of Democratic Party Organization, 1940–1980

ALAN WARE



CLARENDON PRESS · OXFORD

*Oxford University Press, Walton Street, Oxford OX2 6DP*
*Oxford New York Toronto*
*Delhi Bombay Calcutta Madras Karachi*
*Petaling Jaya Singapore Hong Kong Tokyo*
*Nairobi Dar es Salaam Cape Town*
*Melbourne Auckland*
*and associated companies in*
*Berlin Ibadan*

*Oxford is a trade mark of Oxford University Press*

*Published in the United States*
*by Oxford University Press, New York*



*First published 1985*
*First issued in paperback 1988*



*British Library Cataloguing in Publication Data*
*Ware, Alan*
*The breakdown of Democratic Party organization, 1940–1980.*
*1. United States. Political parties: Democratic party, 1940–1980.*
*I. Title*
*324.2736'09*
*ISBN 0–19–827333–9*

*Library of Congress Cataloging in Publication Data*
*Data available*

*Printed and bound in*
*Great Britain by Biddles Ltd,*
*Guildford and King's Lynn*

For Matti and Iain

# Preface

This book is unashamedly old-fashioned. The author is a political scientist who became intrigued by one of the most extraordinary transformations of a political party this century. Unfortunately, as with many institutions which keep few records of their activities, a study of the recent history of America's parties entails the use of a wide range of fragmentary information. In tackling this subject I deliberately avoided the modern American fashion of relying primarily on quantifiable data. The reason for this was quite simple. There were very little data of this kind available, so that a researcher had to choose between examining an extremely narrow range of issues connected with American party decline, or returning to the older tradition of scholarship in which all available sources of information were used. As it was quite clear that the former approach could never help us to understand fully the forces which produced the transformation of the parties, I had no hesitation in opting for the latter. In that sense this book is an attempt to return to a more pluralistic approach to the study of political institutions.

A great many people, and several institutions have made important contributions to this study. Without their help I would not have been able to undertake it, let alone complete it. Unfortunately, in a brief preface I can thank only some of the people to whom I am indebted, and I hope that the others will understand this.

In 1978-9 financial support for the project in Denver and the East Bay was provided by a grant from the Social Science Research Council. In 1981-2 similar support for the project in New York was supplied by the Nuffield Foundation. I am grateful to these two organizations for making the research possible. On both occasions I was able to apply for these grants only because of the system of sabbatical leave operated by my university. I hope all three institutions will find the results worth while.

For it to be successful, field research conducted thousands of miles from home requires the co-operation of others. I was extremely fortunate in receiving generous help from Arthur Shartsis and Professor Nelson Polsby in the East Bay, and from Professor Spencer Wellhofer and Professor John and Cherrie Grove in Denver. My research was also facilitated by the Institute of Governmental Studies at the University of California which made me a Visiting Fellow.

Some of the material in the book was presented in a paper given to the Political Theory Workshop at the University of Warwick, and other material was presented at the Annual Conference of the American Politics Group in 1984. I wish to thank the participants for their useful comments. A number of chapters were read by Iain McLean, Andrew Reeve, and Matti Zelin, and I am grateful for their incisive, scholarly, remarks.

My final debt is to my family. My parents were always supportive of my academic pursuits, and I was sad that my father, James Ware, did not live to see the completion of the project. Jane Ware and Dave Walker were of great help during the early months of the research. The encouragement of my wife Matti has been even more important than her comments on my work. Our son, Iain, taught me the value of organizing my working hours efficiently and has given me tremendous joy during my non-working hours.

*University of Warwick, Coventry* ALAN WARE
*October 1984*

# Note to the Paperback Edition

In the three and a half years that have elapsed since I completed the final draft of this book several important studies have been published which have some bearing on the arguments I was developing then. Of particular significance are the books by Cotter *et al.*, Epstein, and Mayhew.[1] Had I completed my manuscript after these studies were published, there are some respects in which I may have given a slightly different emphasis in the way I presented my own arguments to take account of them. While I do not wish to change any of the conclusions I drew in my study, it may be useful to explain how some of their arguments about, and interpretations of, American parties relate to mine. In particular, I will draw attention to areas where our arguments seem to conflict, or to point in rather different directions. Since the remainder of this note is about disagreement, I wish to emphasize that I believe these scholarly differences relate to rather narrow issues and that the common ground between myself and the other authors is extensive. Furthermore, I would advise anyone (student, political scientist, or someone with a general interest in party politics) to read these other studies – partly because they are examples of fine scholarship, and partly because they expand the debate about American parties into areas which my book does not cover at all.

These works raise three questions to which some attention should be given: (1) Have American party organizations actually declined? (2) Do my case studies cover, as I claim, instances of a previously strong party organization, a previously weak party organization, and an intermediate case? (3) How are American parties to be understood for the purpose of comparing them with parties in other liberal democracies?

The first question is prompted by the arguments of Cotter and his colleagues, which seek to show that the organizational strength of parties, especially at the state level, has actually increased. This book is an expanded version of some of the research they published earlier as articles, and which I mention in Chapter 1, although they also include some data on the strength of local party organizations. Their findings are significant because they demonstrate that, in the case of many state parties, there have been increases in full-time staff and party budgets (though not if this is calculated in constant dollars) since the 1960s, and that many state parties are now able to provide candidates with campaign services. Two preliminary points may be made about their conclusions. One is that the authors themselves argue that these developments are less pronounced among Democratic than Republican parties, and the other is that, by relying on interviews with party officials, Cotter *et al.* may merely have reported *official* versions of what has been happening in state parties.[2] However, my main objection to their arguments is the point I make in note 22 of the first chapter: the data they have gathered are a wholly inadequate test of *party organizational strength.* In making this claim I am certainly not intending to belittle their findings, which, as I have said, are important in understanding the transformation of Democratic party organizations. But the greater institutionalization of state parties which they have discovered is only one, rather small, element in assessing organizational strength. To put the matter rather crudely, a strong party organization is one which, at the very least, can determine who will be the party's candidates, can decide (broadly) the issues on which electoral campaigns will be fought by its candidates, contributes the 'lion's share' of resources to the candidates' election campaigns, and has influence over appointments made by elected public officials. The measures of organizational strength constructed by Cotter and his colleagues are much too narrowly focused to justify their claims that at both state and local levels American party organizations are not in decline.[3]

Obviously, in comparison with many European parties American party organizations have not been very strong, but the evidence which I present in this book is that, at the local level, urban parties in the US have actually become much weaker in a variety of ways. For example, declining influence over the nomination of candidates, and a much-reduced capacity to supply campaign resources (compared with those which individual candidates can supply), have affected considerably the balance between party organizations and individual candidates. To study party organizations, as Cotter *et al.* do, without looking at them in relation to candidates is to disregard the enormous effect the latter have had on the transformation of the former.

Almost certainly, one effect of Democratic party collapse at the local level has been that state party organizations have had to become more institutionalized in an attempt to compensate for this, and indeed institutionalization at this level is precisely what Cotter and his colleagues have discovered. But to describe the developments they have measured as constituting greater organizational strength, at least in the case of the Democrats, is misleading. For example, state parties are even less able to control nominations to state office than they were two or three decades ago, and the campaign funds supplied by individual candidates have grown enormously in this period. At the same time state party budgets have declined in real terms, and party campaign efforts have been directed largely at activities, like 'get out the vote', which candidates are content to leave outside their control. Candidates are willing to see such activities reside within party structures partly because they do not threaten candidate autonomy, and partly, it might be argued, because 'get out the vote' campaigns have the features of a Chicken supergame. As McLean has said of Chicken in relation to lobbying by 'privileged' groups: 'When Chicken is played repeatedly . . . it turns into a game of bluff and precommitment. If there is more than one player who would pay the full cost of the lobby if he had to, each of them has an incentive to precommit himself to not paying, and thus force the other(s) to pay.'[4] Often the mobilization of loyal party electorates does involve a 'privileged' group, in that well-funded individual candidates for major offices might well be able to provide this service on their own. But they also have an incentive to force similarly placed candidates to incur the cost of doing so, and this would likely get them all into a Chicken supergame. One way of avoiding the expense and uncertainty of this is to cede campaign activities in which there is a large public-good element to party organizations.

In other words, far from demonstrating that party organizations have become stronger in recent years, Cotter *et al.*'s evidence must be considered alongside other available evidence of change in American parties; together they point unmistakably in the direction of two conclusions. First, for all their greater staffing and campaign support at the state and national levels, party organizations have less sway over candidates than they used to. Second, consolidation of the parties at the state level constitutes an appropriate rationalization of party activities in a world in which organizational capabilities at the local level have declined considerably.

The second question, whether my study actually covers instances of both weak and strong party organizations as well as an intermediate case, arises because of Mayhew's very different interpretation of the history of party organization in Denver. Mayhew argues that parties in Colorado have 'a tangibility that parties in most Western states lack, but there is no reason to suppose that traditional organizations have operated in Colorado in modern times'.[5] If he were correct, my claim to have studied a genuinely 'intermediate' case of party organization would be undermined to some degree. On his interpretation, the Democratic party in Denver would always have had much more in common with weak party organizations than I suggest it does.

However, Mayhew does not mention at all the Stapleton machine in Denver, which dominated Denver municipal politics between 1923 and 1947, and which I discuss on pp. 63-5. Indeed, the party's activities in the 1950s, which Mayhew does discuss, can be understood fully only against the background of the collapse of the Stapleton machine. Although it was a personal machine, it had strong connections with the Democratic party organization in the city, and the liberal wing of the party could not be effective in municipal politics until Stapleton's electoral defeat in 1947. By excluding cases of machine politics, like that of Stapleton, which were instances of 'quasi-party-machines', Mayhew underestimates the indirect patronage available to some party leaderships in earlier years. By the 1950s the Denver county party's power rested almost exclusively in its power to determine the nomination in partisan elections and in its ability to organize electioneering in general elections. In this respect it was very different from the weak California parties of the same era, but equally it was distinct from the patronage-dominated, hierarchical party organizations still found in the New York boroughs of that period.

The third question – how we are to understand American parties in relation to parties elsewhere – reveals a difference of emphasis between Epstein and myself in our accounts of American parties. While I see American party machines as having been a variation on the caucus-cadre type of party, Epstein argues that

> The machines themselves may look like variants of cadre parties because their bosses operate through the same familiar committee structures. But they were significantly different, it seems to me, as long as they had large numbers of regularly organized party workers, thanks to bountiful patronage jobs, and thus an organizational substance that cadre parties by definition do not have.[6]

This conclusion follows from Epstein identifying cadre parties with 'skeletal organizations' (organizations 'without organized rank and file'), so that for him it is the *extent* of the party organization which distinguishes post-machine American parties from most European parties. However, in describing American parties as examples of caucus-cadre parties (in Chapter 1), I was concerned with the *principle* of organization: American parties have featured a decentralization of power to local élites and have lacked the formal sharing of power and costs among activists found in mass-membership parties.

Rather more than a mere difference in the use of terminology between Epstein and myself is involved here. While Epstein tends to draw a contrast between American parties and 'the dues-paying mass membership parties characteristic of Europe', albeit a contrast which he believes has become less evident in the last thirty years,[7] my own approach points to a less clear-cut distinction between the two continents. Even during this century there have been many parties in Europe which have more closely resembled the American model, in which control rests with relatively small élites, than the early Catholic and socialist exemplars of mass parties. As von Beyme notes, in twentieth-century Europe 'an effective party organization could co-exist with an ideological rejection of the modern form of organisation'.[8] Again, parties based on the patronage of local élites were able to survive in some European countries, especially those (such as Italy) with large peasant populations, well into the twentieth century. Moreover, when they did feel pressure from the organizational success of mass-membership parties, caucus-cadre parties often developed 'hybrid' forms of organization that went only some of the way towards taking on the form of a mass-membership party.[9] While they differ from the American parties, the non-Gaullist centre-right parties in France, the British Conservative party, and the Italian Christian Democrats are also very different from a 'pure' membership-party model. By understanding the similarities between some of these European parties and American parties we can, perhaps, best appreciate the truly unusual features of the latter.

*University of Warwick, Coventry* ALAN WARE
*March 1988*

## Notes

[1] Cornelius P. Cotter, James L. Gibson, John F. Bibby, and Robert J. Huckshorn, *Party Organizations in American Politics* (New York: Praeger, 1984); Leon D. Epstein, *Political Parties in American Mold* (Madison: University of Wisconsin Press, 1986); and David R. Mayhew, *Placing Parties in American Politics: Organization, Electoral Settings, and Government Activity in the Twentieth Century* (Princeton, NJ: Princeton University Press, 1986).

[2] This second argument is outlined by Pamela J. Edwards, 'State Party Organization Revitalization: A Critical Assessment', paper prepared for the 1984 meeting of the Southern Political Science Association, and is cited by Epstein, *Political Parties in the American Mold*, p. 152.

[3] Cotter *et al., Party Organizations in American Politics*, pp. 31–4, 57.

[4] Iain McLean, *Public Choice* (Oxford and New York: Basil Blackwell, 1987), pp. 67–8. The term 'privileged groups' is Olson's, and refers to groups which contain at least one member who gains enough from a public good to supply it (if necessary) himself. Mancur Olson, *The Logic of Collective Action* (Cambridge, Mass.; Harvard University Press, 1965) and *The Rise and Decline of Nations* (New Haven: Yale University Press, 1982).

[5] Mayhew, *Placing Parties in American Politics*, p. 176.

[6] Epstein, *Political Parties in the American Mold*, p. 144.

[7] Ibid., pp. 4, 144–6.

[8] Klaus von Beyme, *Political Parties in Western Democracies* (Aldershot: Gower, 1985), p. 163.

[9] See my account of this in Alan Ware, *Citizens, Parties and the State* (Cambridge: Polity Press and Princeton, NJ: Princeton University Press, 1987).

# Contents

# 1
# Introduction

This book begins with what may seem a wholly unremarkable claim: America's political parties have been in decline. It is unremarkable because the demise of the parties has been proclaimed by political journalists and political scientists for a number of years. If we wished to date the beginning of the era when party decline became generally acknowledged, we might suggest that it occurred between about 1969 and 1972. In the former year Walter Dean Burnham published his article, 'The End of American Party Politics', in *Trans Action*, while the latter year saw the appearance of David Broder's widely read book, *The Party's Over*.[1] Since then the titles of many books and articles, including one by this author, have reflected the view that the parties were disintegrating.[2]

Of course, the establishment of any widely accepted opinion is likely to provoke a reaction, and this instance is no exception. It is probably fair to say that reactions to the view of party decline divide into three categories. There are the self-confessed 'hopefuls' who would like to see the parties revived because they believe that they are crucial to the satisfactory working of a democracy.[3] Then there is a group of scholars whose research has shown that, at both national and state levels, the parties are run more professionally than they were.[4] Finally, it has been shown that within the Republican party in Congress there has been a great increase in centralized fund-raising and in the disbursement of funds to congressional candidates.[5] Few serious objections or counter-arguments can be raised against any of these reactions, for the points their proponents make are well taken. However, it would be highly misleading to suggest that these views provide good grounds for abandoning the general conclusion that the parties have declined.

Yet anyone wishing to express scepticism about the validity of the party-decline hypothesis would have some justification for doing so. While some aspects of the process of decline have been well documented, other equally important ones have been left unexplored. The subject of this book is one of the 'glaring omissions' from the research agenda. Until now few people have addressed the question of how the party organizations have declined in what was their principal base of power, the cities and counties, nor has there been much research into the effect of organizational decline on the conduct of electoral politics at the local level. To understand why these issues are crucial for anyone who purports to explain the transformation in America's parties, it is useful to begin by looking back to the period before party decline was recognized. We consider what an informed observer writing in, say, 1962 or 1963 might have thought were the most significant features of American parties.[6]

## 1. Some Observations on the State of the Parties in 1962 or 1963

We would suggest that the observer would have given particular emphasis to nine features of the parties:

(i) He would have recognized that the formal dispersion of power in the American constitution had served as a barrier to the development of centralized parties at the national level of politics. Such devices as the separation of powers and federalism meant that powerful party organizations could emerge only at the city or county level, though in some instances this had proved sufficient to give control over a state party. The restrictions imposed by the constitution had helped to make America unlike other liberal democracies, in that, in Duverger's terminology, it had not experienced the displacement of its caucus-cadre parties by centralized mass-membership parties.[7] Consequently, the concept of a party member was one which remained ambiguous when applied to American parties. Equally, the distinct legal identity of each state's parties meant that, in some circumstances, it was appropriate to speak of the Democratic *parties* and the Republican *parties* in America.

(ii) Our observer would have pointed out that, although America's major parties were all of the caucus-cadre type, there were many varieties of organizational practice to be found in its city and state parties. At one extreme were several western states, of which California was the most conspicuous example, in which party organizations had been weak since the early decades of the century. In California the organizations' leaders had little control over the choice of a party's candidates for any kind of elected office. The central actor in the nomination process was the candidate himself: he was responsible for constructing an organizational and electoral coalition to enable him to win the party's primary election. After winning the primary, the candidate would run a campaign for the general election which was independent of his fellow candidates on the party ticket. In other words, party seemed to be little more than a label and, with all local government elections in California being legally non-partisan, it did not exist even as a label at the lowest levels of politics. Towards the other extreme were states such as New York. For any prospective candidate in New York support from the urban bosses was usually the key to success in the nomination process. For the more important offices, such as that of state governor or US Senator, the nomination was decided in a party convention at which the bargaining power of the largest county parties was decisive. However, for lesser elected public offices, virtually all of which were filled by partisan elections, the party's campaign resources at the primaries had normally been too great for the would-be insurgent candidate. At general elections the candidate would campaign as a member of a party ticket even though, especially in the case of major offices, he would also distribute his own literature to the electorate. Between the two extremes of unorganized parties and hierarchically organized ones there were numerous ways in which the party activists, organization leaders, and candidates could interact with each other; each state was different, and the variations in the relationships reflected differences in electoral laws, history, and political culture.

(iii) The observer would have noted that party competition in the states had been increasing since the 1930s. The electoral realignment associated with the New Deal had ended many of the state electoral monopolies, which Schattschneider had called the 'system of 1896'.[8] Under this system party competition had been eliminated in most states, with the second party being uncompetitive in elections for state legislatures and governorships. By the early 1960s over half the non-southern states could be classified as competitive, though in others change was slow, and in the South the Republican parties retained minor-party, and not just second-party, status.[9]

(iv) The observer would know that party identification was the single most important factor in explaining the voting behaviour of individual citizens. An increasing number of studies, especially those of the 'Michigan School', had demonstrated that this, rather than ideology, political issues, or candidate style, was the key variable.[10] While these other factors did affect voter choice, and thus could explain differences in party fortunes at successive elections, electoral politics was essentially party politics.

(v) Although it was still common to speak of 'boss-dominated' parties and 'political machines' in some of the eastern and mid-western cities, our observer would have recognized that most machines were in decline. With the odd exception, including Chicago where a hierarchical organization based on patronage had been rebuilt in the 1950s, few city parties were as powerful as they had been fifty years earlier. A number of explanations had been given to account for the collapse of the exchange process by which party workers received jobs in local government and, in return, performed services connected with electoral mobilization. Among the more important of these explanations were: the growth of high-wage, 'full employment' economy after the Second World War; the extension of the civil service 'merit' system at the expense of patronage positions in state and local government; the emergence, after the 1930s, of a quasi-welfare state which provided the first effective competition for the informal services provided by the urban machines; and the migration of white ethnic voters to suburban counties, and their replacement in centre-city areas by Black, Puerto Rican, and other non-white groups who had little experience in the bargaining politics of the machines.

(vi) The observer would have been aware of a relatively new development: the increasing involvement of issue-oriented activists in the two parties. This was first observable in the Democratic party in the 1950s, when young liberal activists either organized themselves into political clubs which were affiliated to the Democratic party or entered the party organization directly.[11] In both cases their ultimate purpose was the same – to nominate liberal candidates, rather than the more conservative ones who would otherwise be selected. In the Republican party conservative activists in the 'grass roots' organizations were given a major impetus to regroup because of the failure of Richard Nixon in the 1960 presidential election. The conservatives had suffered a major defeat over the Eisenhower nomination in 1952, and by 1960 there was no representative from the right contesting the nomination. After the 1960 election there was a concerted effort by conservatives to make the party more ideologically pure; they took control of many county party organizations, and made possible the nomination of Barry Goldwater as the party's presidential candidate in 1964.[12]

(vii) It would have been obvious that some candidates for governorships and US Senate seats in the larger states had started to make use of campaign consultants and television advertising in their campaigns.[13] However, the observer may well not have foreseen the effect that the availability of the new technologies would have on the parties in the future. After all, contests for the US Senate had been more independent of party campaigns than contests for other offices ever since the first popular elections for these seats in 1914. Furthermore, by the early 1960s no gubernatorial candidate had used television or the other new techniques to campaign independently of his party, except in states such as California where candidate-centred campaigning was long established. Twenty years ago the new techniques merely supplemented the more traditional ones, such as door-to-door canvassing and leafleting, which were used by both parties and candidates.

(viii) While he would have acknowledged that there was considerable variation between legislatures with respect to the degree of party discipline practised within them, the observer would have pointed out that at neither the federal nor the state level was discipline comparable to that found in parliamentary systems. In the Congress, for example, party affiliation was the best predictor of a congressman's vote, but 'breaking ranks' with the party leadership on particular issues was commonplace.[14] While the leadership was not without resources to cajole wayward congressmen, especially junior members, the more serious sanctions had rarely been employed against them since the revolt against Speaker Cannon fifty years earlier. Some legislative chambers, including the US Senate, were even less open to control by party, though an exceptional leader, such as Lyndon Johnson, might use the leadership's bargaining resources to the fullest extent possible.

(ix) In its most public act, that of nominating its presidential candidate, the American party gave to leaders of local and state party organizations what was, in effect, a power of veto. Our observer would have emphasized that the nominating process maintained a balance between a popular component and an organizational component. In a few key primaries the candidate could seek to demonstrate strong voter support, but his prospects for nomination depended on bargaining with party leaders – in both the non-primary states and in many which did hold primaries. As Estes Kefauver had discovered in 1952, a demonstration of electoral appeal could be insufficient to win the nomination. The balance between the two components had been maintained since the early 1920s when the parties had first adjusted to the advent of the presidential primary as a nomination device.

Although some of these nine factors reflected aspects of change in America's parties, essentially our observer would have been reporting a situation of stability and gradual change, rather than a rapid transformation. What textbooks had to say about the parties in 1962 was not that different from what their counterparts had written twenty years earlier.[15] Only two new elements, the rise of the amateur activists and the new campaign technology, would have appeared in the books of 1962, although the continuing decline of the machines and of one-partyism would have warranted more attention than in the books of 1942. The contrast between the points our observer would have made in 1962 and those which would be made by someone performing a similar task today is striking, the latter could scarcely avoid giving particular emphasis to the great changes which have occurred in party politics in the last two decades. The six most conspicuous changes have been the following: the decline of organizational control over nominations to state and local elected offices in places where earlier it had been considerable; the decline in the relative importance of party identification in determining a voter's electoral choices; the rise, in the period 1964–1974, of issue-oriented activism which was only loosely associated with the Democratic party; the spread in the use of the new campaign technology and of campaigning by candidates independently of their party's campaign; the decentralization of power in Congress; and the transformation of the presidential nominating process so that control by party organizations was minimized. Between them, these changes form a major component of what Anthony King has called 'The New American Political System'.[16]

## 2. Political Science and Party Decline

Today, if an observer was to report not merely on the parties themselves but on political science research on American parties, two further points would surely strike him. The first is that, for all the discussion about party decline, most research has focused on single, supposedly discrete areas of that decline. Roughly speaking, in order of their visibility, the topics to which attention has been given are: the decline of party as an influence on voting behaviour, the reform of the presidential nomination procedures, and the fragmentation of power, and hence of party influence, within Congress. Only recently has the first major study of the new campaign technologies appeared and, for all the talk about the decline of the party organizations, there has been little published research on the subject.[17] The second point is that discussions of party change and decline have prompted remarkably little re-examination of the nature of American parties. It is remarkable because we might expect that a period of transformation would generate far more analyses of the concept of party and theories of party development and decay than a period of stability. Reflection prompts two questions: why has a rejection of 'overarching theory' in relation to American parties become so well established, and why does this inhibit our understanding of these parties?

The answer to the former question lies in the development of political science since the Second World War. Between the late 1940s and the 1970s political science in America isolated itself from intellectual movements in western Europe, and was dominated by efforts to create what was thought to be a genuine science of politics. Moreover, at the beginning of this era, when there was a huge increase in the number of political scientists and in the amount of research undertaken, theorizing about political parties became controversial. Despite its considerable merits, especially on the subject of the means by which intra-party democracy could be achieved, the APSA report in 1950 on political parties was generally discredited within the profession.[18] Consequently, researchers on parties tended to steer away from the kind of 'grand theorizing' which the debate about the potential for party government had rendered unfashionable. In turn, this contributed to the failure of Duverger's analysis of party structure in *Political Parties* to spark off the kind of debate in American political science which might have been expected following the publication of a study of its scope. Instead, there emerged in the discipline what is best described as the model of the 'unholy trinity', a device which avoided theorizing about American political parties. In the 'trinity' model there are three distinct elements of the concept of party: the party-in-the-electorate (voting predispositions and behaviour), party organization, and party-in-government (candidates, and relations between elected officials bearing the same party label). Those who adhere to this model do not merely accept that research, say, into party organizations can be conducted without equal consideration being given to voting behaviour and party relations in a legislature, they proceed as if party organizations can be analysed without reference to the more general concept of party. In other words, the concept of party is effectively dissolved, though often those who make use of the model continue to speak and write in an ambiguous way about 'parties'. Of course, those who accept the model may be correct in believing that American parties can be conceived as three components that are separable in political analysis. But this involves a justification, and that requires that it be shown how American parties differ from their European counterparts, for no one has ever suggested that the 'trinity' model is appropriate when discussing European parties.[19] Cursory discussion about the interrelationships of the three components are no more adequate than would be claims about the relationships between four elements (earth, air, fire, and water) in an analysis of the physical world based on this classification.[20] Obviously, fire and air interact, but what the sceptic is entitled to know is why it is appropriate to make use of these distinctions – that 'the world appears to divide up in this way' is an unsatisfactory reply. (If any attempt was made to justify the 'trinity' model, one of the few studies which might be used as a basis for this is Epstein's *Political Parties in Western Democracies*.[21] Although his claim, that the only fundamental function which parties have is structuring the vote, is unconvincing, this is one of the few attempts to provide an alternative framework comparable in scope to that of Duverger.)

The widespread acceptance of the 'trinity' model has been bolstered by the main development in post-war American political science – the rise of behaviouralism and its aftermath. The behavioural movement of the 1950s and 1960s became a war between one conception of political science and its opponents. At their worst, proponents of behaviouralism treated other approaches as unworthy branches of the discipline, and in some university departments political philosophers and others were virtually 'written out' of the profession. Since then, the evangelical proselytizing by behaviouralists has diminished; for example, articles on political philosophy are now published regularly in the *American Political Science Review*. Yet, if political theorists and area specialists are left to 'get on with their own thing', the post-behavioural world of American political science is far from being a melting-pot of different approaches. One legacy of the behavioural movement is that much of the study of American politics, excluding such unpromising areas as the Presidency, has been devoted almost exclusively to the study of problems for which quantifiable data can be obtained. For subjects such as political parties, the old tradition of the political scientist as contemporary historian is largely dead. With the exception of undergraduate textbooks, the literature on American parties is dominated by efforts to quantify, even when this means that trivial research is conducted at the expense of more interesting subjects. Nevertheless, it cannot be denied that most of the interesting questions about the operation of party organizations in America are not amenable to these techniques, and as a result little research has been undertaken on these issues.

There are three main reasons why techniques involving quantification are inappropriate. First, when analyzing the operation and development of any organization, quantifiable data can, at best, illuminate only some of the issues; no student of the British Cabinet would suggest seriously that Mackintosh's classic work could be replaced by a study relying on quantifiable data, nor would he argue that a comparative study of cabinets could be based primarily on such data.[22] Secondly, even those aspects of party organization which could be illuminated by the data pose a problem for the researcher, because so little material of this sort was collected by our predecessors; this means that anyone who wishes to trace, say, the decline in recruitment to party organizations will find hardly any material with which to compare profiles of contemporary organizations.[23] Thirdly, even when quantifiable material exists, it may be of little use because it is unreliable. For example, comparing expenditures in congressional and state elections on the basis of official records is worth while only since the passage of the campaign reform laws in the 1970s. So many candidates evaded the earlier federal and state regulations that the records of expenditures are highly misleading. (In one interview for this study an ex-congressman claimed that frequently he appeared to be one of the highest spending campaigners, but this was because he was one of the few who did not conceal any expenditures.)

A defender of the 'trinity' model, however, might acknowledge that there are problems in conducting research on party organizations, but then go on to argue that this makes little difference to our understanding of American party politics. On this view, we could simply infer from newspaper reports what has been happening to the organizations and, in defending itself, the profession could point to the high quality of research on voting behaviour and on the reform of the presidential nominating process. If resources have been concentrated into only a few issues, at least these have been ones of central importance. Yet, for three related reasons, this defence of the model is inadequate. The first is that the extent of our knowledge of how the parties have declined, and of how competition for lesser offices has been affected by changes in electoral behaviour and in modes of competition for major offices, is limited. Ticket-splitting and candidate-centred campaigning are features of the most visible offices, rather than the lesser ones; but, to the extent that they have contributed to the demise of institutions and techniques that benefit competitors for lesser officers, these factors also alter electoral competition at the lower levels. Unless we examine the interaction of organizations, candidates, and activists at the lower levels of politics, our understanding of the nature and effects of recent developments can only be a partial one.

A second reason for rejecting the defence of the model is that the compartmentalization of party research into discrete areas, and the absence of research on the decline of local and state party organizations, has contributed to the growth of what may be called a 'myth of party decline'. This is the view that, increasingly, party organizations experienced a decline in campaign-related resources so that they were unable to perform electoral functions; individual candidates then started to purchase the resources for themselves. That is, first there was organizational decline, and then new methods of contesting elections developed to fill the void. One example of this kind of view is Sabato's claim about the relationship between the rise of the political consultancy industry and the collapse of the parties: 'The atrophy of the political party system in the United States can hardly be blamed on consultants, often the consultants have more been beneficiaries of the web of forces that have brought the parties so low.'[24] Suspicions about the validity of this sort of claim seem proper when we remember that the ineffectiveness of the parties was not apparent to observers in the early 1960s; but if the parties *were* becoming incapable of performing campaign functions, why was it not apparent? The flaw in the 'myth of party decline' is the assumption that a new style of campaigning, centred on individual candidates, would come into widespread use only if the parties were ceasing to do what once they had done efficiently. To assume this is rather like assuming that railways in nineteenth-century Britain would develop only if the canal system was in decay. The displacement of one industry by another, though, does not occur only when the one becomes inefficient: it can also come about because the new industry can supply a far better product than the old one, whatever the level of efficiency in the latter. At least, it can do so providing there are not barriers to its entering into competition with the older industry. In the case of political parties, the 'barriers to entry' are crucial in explaining the growth of the consultancy industry. In countries such as Britain there are effective barriers to consultants plying their wares to individual candidates, of which the most important, perhaps, are the very low levels of permitted campaign expenditures. Consultants can sell their wares only to the national parties. Thus, despite the fact that membership of British political parties has declined spectacularly in the last twenty or thirty years, electoral competition continued as party-centred competition. In America the absence of these barriers meant that increasingly, from the 1950s onwards, candidates could choose whether to obtain campaign services from party organizations or from consultants. Declining recruitment and organizational collapse may have made it less easy for the parties to withstand the new competition, but they can be said to have brought it about only if it can be shown that the services which the parties used to provide would have been more attractive than those the consultants can supply. The role of the 'trinity' model in contributing to the acceptance of this 'myth' relates to the bias in the direction of the research which the model has fostered. Little research has been conducted as to how candidates choose the strategies they use and whence they get the services to carry out these strategies. It is research that is only partly open to quantification, and it falls outside the discrete areas of activity which characterize the 'trinity' model. For this is a model of politics which is concerned with activity, rather than with the relationships between perceptions, values, and activities, and it is concerned primarily with activity within each of the three 'parts' of the parties.

The model's role in drawing attention away from certain crucial aspects of party decline, and thereby making it possible for a simplified view of the process (the 'myth') to become widely accepted was our second reason for rejecting it. There is, however, a more general reason for this rejection: we would suggest that the widespread use of the model has resulted in the most crucial elements of party collapse remaining unexamined. For all the talk by its advocates about the relationships between the components of the trinity, the model has had the consequence of directing attention away from certain fundamental relationships in electoral and public policy processes. Research tends to be concentrated on voters, or on candidates and public officials, or to a lesser extent on party organizations, rather than on their interaction. This has produced gaps in the research agenda which correspond to key aspects of party collapse. One example is cited here to illustrate the point. One of the more obvious and important hypotheses about party decline is that a reduction in both the level of identification with a party and the propensity to vote for only its candidates is linked directly to reduced party contact with the voter, relative to that of candidate contact. (Contact is taken to include exposure to campaign workers and literature, as well as to candidates and party names on radio and television advertisements.) That is, it seems plausible to argue that a decline in the visibility of party reduces voter loyalty, and in turn this induces candidates to value even less their electoral connection with the party. This mechanism would seem to be at the centre of the process of party decline in America. Yet no major research project has been set up to show the precise link between changes in contact with the voter and the choices he makes between party loyalty and support for individual candidates. We have no idea whether disloyalty increases incrementally or whether it is a process involving 'catastrophes'.

## 3. The Object of this Study

This book came to be written only because of the effect that the 'trinity' model, and trends in American political science, have had on research in America. It began after we attempted to find an answer in the existing literature to a question which seemed both straightforward and important: if, as is generally accepted, party organizations have declined, how has this happened, and how have the relationships between organizations, candidates, voters, and other actors in electoral campaigns been affected, especially at the local level of politics? Having failed to find an answer, we set out to conduct our own research on the subject. The aim of the book, then, is to present some of the evidence we uncovered about party decline, to explain how local party organization collapse came about and how it is related to the rise of the candidate as the fulcrum of the electoral process, and to examine the consequences for electoral politics in America. But, in addition, there is a secondary objective. We hope to refocus attention on *party*, rather than, as in Sorauf's words, on some 'tripartite system of interactions', because ultimately the significance of recent changes in America can only be understood in the context of what is, and what is not, party politics. We can say when party decline becomes an 'end of party politics' only if we have that understanding, and it is that which the a-theoretical approach dominant in America has failed to provide.

Because our research has its origins in an empirical question, and not a conceptual or theoretical one, and because the resulting research has been empirical in nature, we can offer no new analysis of the concept of party with which to begin this study. Despite its obvious limitations, the best starting-point is still Duverger's *Political Parties*. Even if we reject his central thesis about party structure – that the mass-membership parties were an advanced form of party – Duverger's two concepts of the older form of party, the cadre and caucus parties, are useful ones for our purpose. Duverger begins by distinguishing party organization from party membership. He argues that party organizations are composed of 'basic elements'; that is, a party is a 'union of small groups dispersed throughout the country', and he identifies four types of these elements. One of the types is the caucus, or closed group, comprising a small number of members who do not seek to expand their number; this was a device which Duverger believed was appropriate only in the early stages of democratization.[25] In contrast to the more modern 'branch', the caucus is a decentralized form of organizational element.

With respect to membership, Duverger makes a further distinction: that between mass parties and cadre parties. The former seek to spread the various costs of election campaigning among as many people as possible, by admitting dues-paying members. The latter is a 'grouping of notabilities for the preparation of elections, conducting campaigns and maintaining contact with the candidates'.[26] Having explained the differences between types of party in respect of both organization and membership, Duverger then argues that 'cadre parties correspond to the caucus parties, decentralized and weakly knit; mass parties to parties based on branches, more centralized and more firmly knit'.[27] Caucus-cadre parties, therefore, are groups of individuals at local levels of politics who are separate from the party's candidates, and who engage in organizing electoral activity; they are not dependent on any nationally-based organization, nor do they recruit fee-paying members. Branch-membership parties are centrally organized and recruit members in a formal way in order to reduce the cost to each of them of the campaign activities. Yet, if it is a simple form, the caucus-cadre type is still a party – there is something (traditional loyalties, shared patronage, shared political ideas, or whatever) which ties the members of the notability to each other. The value of the Duverger classification is that it reflects two contrasting ways in which collective goods can be pursued: through informal co-operation, possibly supplemented by the 'buying-in' of extra help, and through a formal agreement to share the costs of providing them. Thus, the distinction between the types of party is similar to the distinction between informal, but regular co-operation, between economic actors (such as neighbouring farmers) and the operation of an economic unit as a joint-stock company.

However, several American academics, including Austin Ranney, have rejected Duverger's claim that American parties are of the caucus-cadre type.[28] Before proceeding, it is worth considering three possible objections to conceiving American parties in this way. The first is the argument that Duverger ignores the fact that some states have both legally defined parties and those which exist outside the legal framework.[29]

But, whether a group of notables has to comply with state laws regarding its structure does not mean that co-operation ceases to be informal; the state is merely requiring those who wish to co-operate to conduct part of their operations in a particular way. It is rather like neighbouring farmers being required to state publicly who co-operates with whom, and being prohibited from engaging in restrictive practices which would prevent other neighbours from obtaining help from members of the group. Of course, some of the newer extra-legal parties, especially those in Wisconsin in the 1950s, did establish themselves as mass-membership organizations and do not fit Duverger's model. However, these have never been a common form in America.

A second possible objection is that, even if they did not recruit fee-paying members, the American parties did recruit activists; this was the basis of urban political machines. But to assimilate these activists with dues-paying members is rather like assimilating associate members of the Royal Automobile Club (which is what all but a small minority of 'members' are) with the full members of the London club. The one category contains those who are recruited to perform tasks (or, in the case of the RAC to provide income for the organization) for which they are rewarded in various ways. The other category contains those who are entitled to have a say on the objectives and *modus operandi* of the organization.[30] (Of course, the urban machines differed from the RAC in that faithful contributions to the organizations could eventually be rewarded by recruitment to the notability.) Indeed, it was a conflict over the relative status of the notability and those who were recruited, or had recruited themselves, to perform electoral tasks which led to the Democrats' reforms of their presidential nominating procedures. The 'regulars' did not regard the newer activists as full members of the party, as the notability were, while in turn the newer activists did not accept the role allotted to them.

A third possible objection concerns the effect of introducing a primary election system into a caucus-cadre party, and here it must be admitted that Duverger himself did not fully understand the impact of primaries. He saw primaries as one of the devices which cadre parties could use in seeking to become 'more flexible by pretending to open their ranks to the masses'.[31] But what he failed to see is that in a cadre party primaries can change the relationship between the notables and the candidate. When a candidate can enter a primary without needing the support of the notables, and when he can appeal to the primary electorate directly, then the notability can simply collapse. While in one sense we can say that when this occurs a new notability (the candidates) has arisen, it is important to recognize that this is a very different entity from the old notability. In the extreme case there can develop a style of politics in which the members of the 'new notability' are bound in no way to each other; when this occurs, there can be party-less politics, even though partisan primaries and elections continue to be the format in which politics is organized. In other words, party primaries do not inevitably transform cadre parties into something else, but in some circumstances they can be thus transformed. It was this possibility which Duverger did not comprehend, but it is this which, in effect, is the subject of debate about the 'end of party politics' in America.

To argue that, in general, it does seem appropriate to apply Duverger's conception of the caucus-cadre party to American politics is not to accept his account about the survival in that country of what he regarded as an archaic form. His view was that the American parties did not have to face competition from the supposedly more modern types; equally, it could be argued that the dispersion of power through constitutional arrangements prevents the emergence of any form of party which does not reflect this decentralization. The value of the concept of a caucus-cadre party is that it encapsulates the three main features of American parties which distinguish them from most other parties in contemporary liberal democracies. These are (i) organizational decentralization; (ii) the objective of capturing individual public offices, rather than the broader goal of maximizing influence over government – this is a feature which makes important in America the distinction between candidate and party influentials;[32] and (iii) an organizational style of informal co-operation, rather than the formal sharing of power and costs, though usually this does occur within a well-defined legal structure. Whatever its faults, at least Duverger's construct has the merit that, unlike the 'trinity' model, it does not impose an artificial division on the analysis of American parties. It provides a model of party politics in terms of which we can try to understand the importance of the changes which are grouped under the heading of 'party decline'. We return to consider this model in Chapter 9, after we have examined the various changes which the parties have been undergoing.

## 4. How the Research was Conducted

The obvious problem facing any scholar who sets out to examine the circumstances and consequences of party decline in America would be that of studying a sufficient number of parties so as to provide a complete account. Given the diversity of party practices, there are no 'typical' states or counties which, taken together, would provide a microcosm of the nation. On the other hand, only a few aspects of the changes would be amenable to analysis by the use of aggregate statistical data. In the absence of an army of researchers, the only solution is to reduce the range of issues examined and to employ the 'case study' approach. The disadvantages of this approach are well known, and the researcher is always open to the criticism that he has examined highly unusual instances. Nevertheless, if care is exercised in selecting the cases, the researcher can claim at least that they exemplify particularly important developments and ones which arise in a number of local parties. This was the approach which guided this study. The aim was to provide an intensive study of the parties chosen, but also to select a group which had the potential for some comparative analysis. As a result, it was decided that research would be undertaken on three parties – all of them Democratic, all in urban areas, and all outside the South.

The decision to focus on three parties only was determined by the amount of time involved in examining documents, newspapers, and other printed material and in interviewing individuals who had participated in party activities. Especially in places where there has been little published research into their politics, and where local newspapers are not indexed, the period of preliminary research can be very long. (In the event, nearly two years – between 1976 and 1982 – was spent on field research in the three areas.) That three parties were to be studied also affected the initial decision to focus on only one of the two main parties. While their structures are similar superficially, there are important differences between many Democratic and Republican parties which would affect, especially, inter-state comparisons of the role of party notabilities. For example, most Republican parties are less concerned with intra-party democracy, benefit from more centralized fund-raising, and have always had a much larger proportion of ideological activists.[33] Given these differences between Democratic and Republican parties, there were two main reasons for choosing to examine the former. First, arguments about party decline have been directed more at the Democrats than the Republicans. If we were to understand party decline, it seemed more appropriate to begin by examining the main exemplar of it. Secondly, there was a good case for focusing on urban areas, because these had been the environments most favourable to highly-organized parties in America. However, by the 1960s the Republicans were being reduced to a 'rump' party in many of the largest cities.

The peculiar history of one-partyism at both local and state levels in the South suggested that, whatever the intrinsic interest of party politics in places such as Atlanta or Houston, they would not be good subjects for comparison with northern cities. This consideration further reduced what had originally been a very large number of parties, which could have been included in the study, to a much smaller number. Three additional criteria were then used to shorten the list again – our objective being to select parties which were sufficiently similar to make comparisons worth while, but sufficiently different for a wide-ranging study. These criteria were:

(i) all the parties would be in large conurbations, and would have at least one city with a population of more than 300,000;

(ii) all should have experienced a major insurgence by liberal activists into the party in the 1950s;

(iii) between them the three areas would include an example of what was formerly a relatively centralized patronage-based party, an example of a party which was already organizationally weak before the 1950s, and an intermediate case.

In fact, the choice of the intermediate case was an obvious one, since our interest in the project had started during the completion of a study of intra-party democracy in Denver, Colorado. Before the 1950s the Denver Democratic party had been a hybrid of machine politics and a more open political style; even in the early 1960s the party organization was far more powerful than that usually found west of the Mississippi river. California's East Bay area, which includes the city of Oakland, was an obvious example of weak party organization. While it would have been rejected if an attempt had been made to find a 'typical' area, the peculiar impact that both white and black radicalism had in the East Bay in the 1960s meant that it could reveal in an exaggerated form the effects of issue conflict in that era. Finally New York City was chosen as the third area. The fact that its parties were borough-centred, and not city-centred, meant that there was considerable scope for comparison within the city; this outweighed the obvious advantages of smaller, and less complex cities, such as Boston and Philadelphia.

The main research in the East Bay and Denver was undertaken in 1978 and 1979, while in New York it was conducted between 1981 and 1982. During these periods interviews were held with a total of 135 people, all of whom had been involved in aspects of party politics.

Although the focus of the research was on Democratic parties, several Republican party officials and candidates were included among the interviewees. In addition to the interviews and the examination of documents and newspapers, two mail surveys were also conducted as part of the project. Information about the interviews and the surveys is given in the 'Bibliography and Other Sources' at the end of the book.

The period of party decline which was examined in the study of the three areas was from about 1940 to about 1980, although most attention was devoted to the 1960s and 1970s. The point of taking a forty-year period for the study was that many of the rapid changes which occurred from the 1960s onwards had their origin in social, economic, and political developments of the 1940s and 1950s. The reason for focusing more on the years from 1960s onwards is that, in some respects, our knowledge of the parties is somewhat greater for the earlier years. Political science was not as encumbered then by the constraints of behaviouralism, and scholars did not shy away from examining party organizations just because statistical data could not be obtained. Detailed studies of parties based on original research, such as Key's *Politics, Parties and Pressure Groups* and Wilson's *The Amateur Democrat*, were generally not replicated after the mid-1960s.[34] This book represents an attempt to return to the older tradition.

After Chapter 2, the rest of the book explores the main issues which came out of the research. In Chapter 2 itself we explain how social and economic developments in, and the political institutions of, the three areas relate to those in the rest of the country. In Chapter 3 we examine the state of the parties from 1940 to 1960. While some were wracked by faction, many parties were merely experiencing a quiet decline in membership and were becoming less centralized, and there were some instances of party revival. We suggest that there is no evidence to support claims that the parties generally were becoming incapable of performing electoral tasks. Chapter 4 is concerned with the problem of declining participation in the parties – both by 'professionals' and 'amateurs'. In Chapter 5 we explain that, apart from declining recruitment, there were various other factors which led in the 1960s and 1970s to the parties becoming much weaker both in the nomination process and as electoral intermediaries. In Chapters 6 and 7 we examine the replacements for the party organizations – the individual candidates who can buy their campaign services elsewhere. Chapter 6 focuses on the candidate and the conditions in which he becomes free of party ties, and those which provide him with an incentive to intervene actively in other elections. Chapter 7 examines the use which candidates for lesser offices can make of the new campaign technologies, and also the effects that their use by major office candidates has on lesser office contests. In Chapter 8 we examine the state of the coalitions between the Democratic party and the labour unions and between the party and blacks. Finally, in Chapter 9 we turn to consider the question of whether what has happened to the Democratic party constitutes an 'end of party politics'.

## Notes

[1] Walter Dean Burnham, 'The End of American Party Politics', *Trans Action*, 27 (1969), 12–22; David Broder, *The Party's Over* (New York: Harper and Row, 1972).

[2] For example Ruth K. Scott and R. J. Hrebenar, *Parties in Crisis* (New York: Wiley, 1979); Jeff Fishel (ed.), *Parties and Elections in an Anti-Party Age* (Bloomington, Ind.: Indiana University Press, 1978); Martin P. Wattenberg, *The Decline of American Political Parties, 1952–1980* (Cambridge, Mass. and London: Harvard University Press, 1984); Alan Ware, 'The End of Party Politics? Activist-Officeseeker Relationships in the Colorado Democratic Party', *British Journal of Political Science*, 9 (1979), 237–50.

[3] Gerald M. Pomper (ed.), *Party Renewal in America* (New York: Praeger, 1980).

[4] Cornelius P. Cotter and John F. Bibby, 'Institutional Development of Parties and the Thesis of Party Decline', *Political Science Quarterly*, 95 (1980), 1–28, and Cornelius P. Cotter, James L. Gibson, John F. Bibby, and Robert J. Huckshorn, 'State Party Organizations and the Thesis of Party Decline', paper presented at the Annual Meeting of the American Political Science Association, Washington, DC, 1980.

[5] *Congressional Quarterly Weekly Reports*, 2 July 1983, 1345–51.

[6] The arguments presented in the following section were first introduced in Alan Ware, 'Party Decline and Party Reform', *Teaching Politics*, 12 (1983),82–96.

[7] Maurice Duverger, *Political Parties*, 2nd English edn. (London: Methuen, 1959).

[8] E. E. Schattschneider, *The Semisovereign People* Hinsdale, Ill.: Dryden Press, 1975), ch. 5.

[9] Jewell and Olson indicate seventeen states as being competitive in elections between 1946 and 1958. In the late 1950s several other states, including California and Wisconsin, could be regarded as two-party competitive. Malcolm E. Jewell and David M. Olson, *American State Political Parties and Elections* (Homewood, Ill.: Dorsey Press, 1978), 32.

[10] Angus Campbell, Philip Converse, Warren Miller and Donald Stokes, *The American Voter* (New York: Wiley, 1960); Angus Campbell, Philip Converse, Warren Miller and Donald Stokes, *Elections and the Political Order* (New York: Wiley, 1960).

[11] See James Q. Wilson, *The Amateur Democrat* (Chicago: University of Chicago Press, 1962), Edward N. Costikyan, *Behind Closed Doors* (New York: Harcourt, Brace, 1966); Francis Carney, *The Rise of the Democratic Clubs in California* (New York: Holt, 1958); and Alan Ware, 'Why Amateur Party Politics has Withered Away: The Club Movement, Party Reform and the Decline of American Party Organizations', *European Journal of Political Research*, 9 (1981), 219-36 for earlier accounts of this development.

[12] On the rise of 'Goldwaterism' see John H. Kessel, *The Goldwater Coalition* (Indianapolis: Bobbs-Merrill 1968), and Aaron B. Wildavsky, 'The Goldwater Phenomenon: Purists, Politicians and the Two-Party System', *Review of Politics*, 27 (1965), 386-413.

[13] The earliest major study of this was Stanley Kelley, *Professional Public Relations and the Political Power* (Baltimore: Johns Hopkins Press 1956).

[14] The most important early study was perhaps Julius Turner, *Party and Constituency* (Baltimore: Johns Hopkins Press, 1951); for an example of a state in which the parties had unusual power see W. Duane Lockard, 'Legislative Politics in Connecticut', *American Political Science Review*, 48 (1954), 166-73.

[15] It is interesting to compare Frederick A. Ogg and P. Norman Ray, *Introduction to American Government*, 7th edn. (New York: Appleton-Century, 1942), chs. 11 and 12 with William H. Young, *Ogg and Ray's Introduction to American Government*, 12th edn. (New York: Appleton-Century, 1962), chs. 8 and 9.

[16] This is the title of King's book. The article on parties in this collection of essays ('The Political Parties: Reform and Decline' by Austin Ranney), like many other recent studies of the parties, focuses entirely on the change in the presidential nominating system. Anthony King (ed.), *The New American Political System* (Washington, DC: American Enterprise Institute, 1978).

[17] Larry J. Sabato, *The Rise of Political Consultants* (New York: Basic Books, 1981).

[18] Committee on Political Parties of the American Political Science Association, 'Toward a More Responsible Two-Party System', *American Political Science Review*, 44 (1950), Supplement. Its immediate critics included Austin Ranney, 'Towards a More Responsible Two-Party System: A Commentary', *American Political Science Review*, 45 (1951), 488-99, and T. William Goodman, 'How much Political Party Centralization do we Want?', *Journal of Politics*, 13 (1951), 536-61. For an assessment of the arguments about intra-party democracy contained in the report, see Alan Ware, *The Logic of Party Democracy* (London: Macmillan, 1979), ch. 5.

[19] Major comparative studies of parties do not use the trinity model: see, for example, Giovanni Sartori, *Parties and Party Systems* (Cambridge: Cambridge University Press, 1976). One aspect of the peculiarity of the model was noted a number of years ago by Anthony King: 'It is common in the United States for writers on parties to refer to "the party in the electorate", sometimes as if it were on a par with the party in Congress or the party organization. The notion of party-in-the-electorate seems a strange one on the face of it. It is rather as though one were to refer not to the buyers of Campbell's soup but to the Campbell-Soup-Company-in-the-Market.' 'Political Parties in Western Democracies', *Polity*, 2 (1969), 114.

[20] An example of the genre, and one we have chosen because it is well known

and an otherwise well-argued book, is Frank J. Sorauf, *Party Politics in America*, 4th edn. (Boston: Little, Brown, 1980), p. 8. The structure of Sorauf's book – including its section headings – is based partly on the trinity model, and his justification for this is both very brief and revealing. He asserts that one of three ways in which political parties can be defined is as a social structure: 'Large organizations or social structures are people in various roles, responsibilities, patterns of activities, and reciprocal relationships. But which people, what activities, what relationships are we talking about when we speak of the two major American parties?' He lists some possible people who might be included, and then simply claims: 'The major American political parties are in truth three-headed political giants, tripartite systems of interactions, that embrace all these individuals.' Sorauf never questions whether the idea of a political-party-as-a-social-structure is meaningful if the concept of 'social structure' is made so wide as to include voter loyalty to parties. Nor does he justify his claim that the parties are 'three-headed giants'; the reader is told that they are, just as he might be told that the universe is composed of fire, air, earth, and water.

[21] Leon D. Epstein, *Political Parties in Western Democracies* (New York: Praeger, 1968).

[22] John P. Mackintosh, *The British Cabinet* (London: Stevens, 1962). A typical example of the failure to appreciate the limitations of quantitative techniques when examining American party organizations is the work of James L. Gibson *et al.* These researchers have made some interesting and important discoveries about the institutionalization of party organizations, but they seem to believe that the data they obtained demonstrated that the organizational strength of state parties had generally *increased* during the 1960s and 1970s. This counter-intuitive argument was possible because they identified 'organizational strength' with the maintainance of permanently-staffed headquarters and the performance of a number of electoral activities. More parties were professionally staffed by the late 1970s and some had expanded the range of their electoral activities. But these changes are only two elements of what normally can be described in political discourse as 'organizational strength'. What Gibson *et al.* did not show was how both the absolute and relative contributions by state parties to total electoral activity had changed. This kind of information would be extremely hard to quantify though their own data, which show that in real terms the average state party budget declined between 1960 and 1980, would suggest that state party contributions declined considerably. Moreover, they do not seem to realize that the institutionalization of state parties might be linked directly to the decline of county parties; for example, it might well be the case that, even with lower budgets, state parties have become better managed in order to impose some order on party politics when the organizations at the county level have been disintegrating. This sort of hypothesis cannot be examined with their kind of analysis. James L. Gibson, Cornelius P. Cotter, John F. Bibby, and Robert J. Huckshorn, 'Assessing Party Organizational Strength', *American Journal of Political Science*, 27 (1983), 193–222.

[23] Even in the case of Pittsburgh, a city on which a considerable amount of research had been conducted earlier, Weinberg *et al.* had little data on activities with which to compare their own data for the period 1971–6. Lee S. Weinberg, Michael Margolis and David F. Ranck, 'Local Party Organization: From Dis-

aggregation to Disintegration', paper presented at the Annual Meeting of the American Political Science Association, Washington, DC, 1980. Moreover anyone examining the activities of organizational members would probably find survey data inadequate; in Denver, for example, the members *claimed* to be far more active than they appeared to be.

[24] Sabato, *The Rise of Political Consultants*, p. 268.

[25] Duverger, *Political Parties*, p. 20.

[26] Ibid., p. 64.

[27] Ibid., p. 67.

[28] Austin Ranney, *Curing the Mischiefs of Faction* (Berkeley: University of California Press, 1975), p. 148.

[29] Ranney, *Curing the Mischiefs of Faction*, p. 148, cites the arguments of Epstein and Sorauf.

[30] More generally it should be noted that some non-political organizations describe those for whom they provide a service as 'members' without the latter being granted any control over the organizations. Hansmann cites the (American) Book-of-the-Month Club as one example, and he goes on to note that in 'non-profit' organizations the term 'member' can be used 'so broadly as to have little definite meaning'. Henry B. Hansmann, 'The Role of Non-Profit Enterprise', *Yale Law Journal* 89 (1980), 841.

[31] Duverger, *Political Parties*, p. 66.

[32] It is this objective of capturing offices which has been formalized in the analytic approach known as 'ambition theory'. Unlike the trinity model, this does not dissolve the idea of party; it merely assumes that electoral competition can be examined by reference to individual candidates, and not parties. See Joseph A. Schlesinger, *Ambition and Politics* (Chicago: Rand McNally, 1966).

[33] On the differences between Democratic and Republican fund-raising see Xandra Kayden, 'The Nationalizing of the Party System', in Michael J. Malbin (ed.), *Parties, Interest Groups and Campaign Finance Laws* (Washington, DC: American Enterprise Institute, 1980); on the ideological nature of Republican activists, see David Nexon, 'Asymmetry in the Political System: Occasional Activists in the Republican and Democratic Parties', *American Political Science Review*, 65 (1971), 716–30.

[34] V. O. Key, *Politics, Parties and Pressure Groups*, 4th edn. (New York: Crowell, 1958); Wilson, *The Amateur Democrat*.

# 2
# New York, Denver and the East Bay in the National Context

In this chapter we examine the economic, social, and institutional background to the changes faced by the Democratic parties in our study. While some of these developments, practices, and institutions were similar to those elsewhere, others were unusual, and some were unique; the object of the chapter is to allow the reader to see how the three areas fit into the national context. It is directed especially at non-American readers, and those familiar with the areas may wish to omit section 1, for it provides few original insights into the economic and social changes in the period.

## 1. Economic and Social Change, 1940–1980

Throughout the period of industrialization in America, New York was the largest city in the country. By consolidating with three surrounding counties in 1897, the original city of New York (Manhattan) ensured that it would remain part of America's largest city for decades, if not centuries.[1] Through consolidation New York became a city with five constituent boroughs (Manhattan, the Bronx, Queens, Brooklyn, and Staten Island) each of which was a county.[2] In the first four decades of this century its population increased, both absolutely and as a proportion of the nation's population. More than one American in eighteen was a resident of New York City in 1940. Quite simply, consolidation had produced a unit of government unlike any other in America. The city's population increased again between 1940 and 1950, to nearly 7.9m., though by then less than one in nineteen Americans was a New Yorker. Even so, from 1953 to 1962, the city still had twenty-two of the 435 members of the House of Representatives. Apart from New York State itself, only four states had larger delegations than New York City in the Congress: California (30), Pennsylvania (30), Illinois (25), and Ohio (23).

Of course, the importance of New York at the time did not derive merely from its size. A geographically isolated city could never have attained the dominance that New York did, unless it was also the nation's capital. What contributed to New York's pre-eminent position were four inter-related factors. First, it was a major element of the most populous state in the Union: in the 1950s the state had a total of forty-five congressmen. Secondly, it was part of a larger metropolitan area which included the adjacent cities of north-eastern New Jersey; even excluding the 'commuter belt' of western Connecticut, this was an area with 13m. people. It was a major industrial region which contained the commercial and financial capital of the country. Thirdly, it was in the middle of an even larger socio-economic unit, the coastal area of the north-eastern states, stretching from Washington to Boston, which in most respects was the centre of the nation. Fourthly, the average citizen of New York was wealthy in comparison with the average American; in 1940 average per capita income in America was only 58 per cent of that of New York City (Table 2.1). Coupled with its economic dominance, the city had unchallenged leadership in American cultural affairs, so that all it lacked was significance as a governmental centre.

**Table 2.1**: Changes in per capita personal income for the United States and New York City, 1940–1975 in constant dollars

| | United States | New York City | US per capita income as a proportion of NY per capita income |
|---|---|---|---|
| 1940 | 1,410 | 2,422 | 58 per cent |
| 1950 | 2,075 | 3,000 | 69 per cent |
| 1960 | 2,505 | 3,369 | 74 per cent |
| 1965 | 2,947 | 3,722 | 79 per cent |
| 1970 | 3,410 | 4,076 | 83 per cent |
| 1975 | 3,630 | 4,019 | 90 per cent |

*Source:* Bernard R. Gifford, 'New York City and Cosmopolitan Liberalism', *Political Science Quarterly*, 93 (1978–9), 561.

However, the 1950s were a turning-point for New York. Although there was a continuing inflow of Puerto Rican and southern black migrants, for the first time the city's population declined in size. Moreover, if the metropolitan area continued to grow in population and in economic activity, it did so at a much slower rate than other urban regions, particularly those in the western states. The Los Angeles area was emerging as a major rival to New York: in 1940 it had a population of 3m., but by 1960 it had a population of more than 6.7 m. residents – approximately half the size of the New York metropolitan area. Symbolic of this challenge to New York power were the decisions taken in 1958 by the owners of the New York Giants and Brooklyn Dodgers baseball teams to move their teams to new stadia, built especially for them, in California. For the next three seasons there was only one major-league baseball team in New York. California wealth, and the rapidly increasing population there, made such humiliating coups possible. Nevertheless, New York enjoyed two overriding advantages in its coming rivalry with Los Angeles. The latter was geographically isolated – from the nation's capital, from the industrial heartlands of the mid-west and north-east, and from Europe. For example, the European stockmarkets and governments conclude their work before normal business hours begin in California. New York had the further advantage of a leadership position in fields in which tradition or experience were important. Wall Street or the Broadway theatres could not be transferred like a baseball team. What the New York area was losing in the 1950s, and what it would continue to lose, was industrial development and symbolic leadership. Industry could choose to locate elsewhere and coast-to-coast television enabled California to lose its quasi-colonial status. Indeed, with its expertise in the film industry, Los Angeles was a natural centre for television production.

The process of decline continued in the 1960s and 1970s, and in some ways that decline was spectacular, though it is misleading to describe it as a typical example of urban decay. There was a small increase in the population of New York City in the 1960s, before it declined by nearly 800,000 to 7.1m. in the 1970s. By 1980 only one American resident in thirty-three lived in the city. The congressional redistricting which followed the 1980 census reduced the city's delegation in the House from seventeen to thirteen; its representation was no longer comparable to that of a large state. Apart from New York State, eight states have larger delegations in the 1980s. There was a corresponding decline (of one million) in the population of the New York metropolitan area, while the population in the Los Angeles area was increasing so that it was about three-quarters the size of New York by 1980. New York's position as the dominant urban area in the country was rapidly being eroded. Of course, a declining population has been a feature of many urban areas in the north-east and mid-west of the country. What is interesting in the case of New York is the speed of the decline, the fact that it occurred without the racial minorities forming a majority of the population, and that it occurred in a city which had become so pre-eminent. To understand what has happened, we must consider New York in the context of more general demographic changes in America.

Since the Second World War the western states have been the beneficiaries of internal migration in America. The populations of the south-western states have also increased disproportionately owing to illegal immigration from Mexico and Latin America. More recently, since the civil rights 'revolution', the southern states have joined the west in being popular destinations for internal migrants. In the 1970s the population of every southern and western state increased by more than the national average. Correspondingly, the population of every north-eastern and mid-western state either decreased, or increased by less than the national average – with three minor exceptions (the small states of Maine, New Hampshire, and Vermont).[3] Two states, New York and Rhode Island, had fewer residents in 1980 than in 1970, and the absolute and relative decrease in New York was much the greater. Thus, the impact of population decline in New York City might seem to make it an extreme case of a general national trend. But is this so? Before making such a claim, it is worth bearing in mind two aspects of demographic change usually associated with urban depopulation. One is suburbanization: the development of virgin land at the edges of cities, often located in different local government jurisdictions. This has reduced and often reversed, population growth in the cities. The other, and related phenomenon, is the increasing proportion of black and other minorities in city-centre populations, a development which is frequently linked to the acceleration of white exodus to the suburbs.

Now New York was different from many other eastern and mid-western cities, in that it continued to be like an immigrant city even after mass foreign immigration ended in the 1920s. It was particularly attractive to southern blacks because of its tradition of welfare-oriented state government. Furthermore, it became the main centre for Puerto Rican immigration. Thus, the city was at its greatest size in relation to the rest of the country not when it was the major port of entry from Europe, but in the 1920s and 1930s when the last European immigrants were having families and when internal migration to the city was at a high level. The further boom in internal migration in the 1960s led to an increase in the city's population in that decade, but this boom could be maintained only if the supply of low-paid jobs was also maintained, or if welfare services remained that much better than in the South or

Puerto Rico. In fact, from the second half of the 1960s onwards, the supply of low-paid, unskilled jobs in the city declined, and the welfare services in other states came more closely to resemble those in New York. The result was that the migration flow reversed; this was not so much symptomatic of a collapse in the city's economy, but rather a change in the nature of that economy. In other words, New York lost a migrant base to its population, something which most other cities had had to a much smaller extent. The loss of this base is reflected in the distribution of population loss in the city. It has been heavily concentrated in places such as the nationally famous South Bronx, a home for those at the bottom of the economic ladder since early this century.

Although the low-paid job sector has declined, economic activity in Manhattan has boomed since the national slump of the early 1970s. Many businesses still want to be located there. Moreover, even with population drift to the suburbs, the very size of the city and the diversity of its neighbourhoods meant that it did not become a city dominated by poor, non-white minorities. Estimates of the size of the non-white population have varied because of the non-congruence of the two relevant questions in the US Census, but at the end of the 1970s that population formed between only 39 per cent and 50 per cent of the city's population. The concentration of this population in particular districts, together with low levels of political participation among the city's minorities, prevented the development of any expectation that non-whites might soon take political control of the city. There was no evidence of a rapid exodus to the suburbs by whites fearful of non-white domination. Indeed, at the end of the 1970s the minorities were even further from power in New York City than they had been a few years earlier.

This is not to deny that non-white groups have increased in size. According to the US Census, their numbers increased by at least 320,000 at the same time as the city's population was falling by over 800,000. But in the 1970s the racial composition of the city was not changing as much as the gross census data for that decade might suggest. There are four reasons for this. First, as was mentioned earlier, the mass flow of non-white migration to the city was ceasing. Secondly, there is evidence that the undercounting of minorities in the census was greater in 1970 than in 1980, so that the increase in these populations over the ten years has been exaggerated. Thirdly, part of the decline in the white population was the result of the children of the post-war baby boom leaving the family unit. This also occurred in the older Nassau County suburbs, adjacent to New York City, and constitutes a non-recurring, if sharp, decline in the population.[4] Fourthly, most virgin land remaining for future development in the New York area is a considerable distance from Manhattan, and this makes it less attractive for potential, mainly white, commuters. The problem of transportation from the suburbs has in turn contributed to an increased demand for residential property in Manhattan. Indeed, the 'gentrification' of many parts of the borough, including previously run-down areas of the West Side, would have proceeded much more quickly but for the city's rent control laws.

By the end of the 1970s New York was experiencing both urban decay and economic expansion. Overall there had been a considerable change in the social composition of the city since 1940, but then this had always happened to the city. The average New Yorker was no longer rich in comparison with the average American, but even in 1975 personal income was still higher in New York than in the country as a whole (Table 2.1). The spectacular evidence of social decay – population loss, the South Bronx, and the city's insolvency in 1975 – must be balanced against other evidence of a slower transformation when examining prospects for future change in party politics in the city. Nevertheless, it cannot be denied that, in the period of our study, these rather exceptional circumstances did affect the changes in the Democratic party's organizations.

If New York was part of a declining north-east, albeit a somewhat unusual case, the Denver region was one of the principal beneficiaries of the westward migration. The number of residents in the Denver metropolitan area doubled in the twenty years after 1940, to 930,000. The city of Denver itself increased its population by over one half in this period, but the post-war years were more significant as years of development in the surrounding suburban counties. By 1960 the city had been largely 'filled in', and thereafter virtually all the population increase was in the peripheral counties. In this respect Denver was similar to many other large western cities. In 1940 approximately three-quarters of metropolitan residents lived in the city, in 1960 a little more than half of them did so, and by 1980 fewer than one resident in three lived there; in 1980 the metropolitan region had a population of 1.6m. By 1960, as in New York, the political balance between city and state, which derives ultimately from relative population sizes, was changing. Denver had been the dominant unit of local government in the state; with 31 per cent of the state's population in 1950, the city had sufficient representation in the legislature to make urban-rural conflicts of central importance in Colorado politics. When the suburban counties became more populous, the nature of this long-standing political cleavage changed, because suburban interests differed from those of Denver. The power of the city in state politics declined after 1960 because of the economic boom in the surrounding region.

Many factors contributed to the rapid economic development, from 1940 onwards, of a metropolitan area located in semi-arid, high-plains land. Only some of them need be mentioned here. One important consideration was that sufficient water could be pumped from the western slopes of the Rocky Mountains. The Rockies not only provided water, but attracted tourists in an era when air and car travel could be afforded by many Americans. The area had also long been a centre of expertise in the extraction of minerals, a sector of the economy which was to be crucial in the economic expansion of the south-west. Again, Denver had become a centre for scientific and technological research with the establishment there of various federal government agencies in the 1930s and 1940s. Finally, being the largest urban region in the vicinity of the Rocky Mountains, Denver was already the regional headquarters and distribution centre for many national firms. Even in 1960 the nearest large metropolitan areas (that is, those with a population of more than three-quarters of a million) were remote; San Francisco-Oakland was 1,200 miles west, Dallas 750 miles south-east, Kansas City 600 miles east, and Minneapolis-St. Paul 850 miles north-east. Expanding economic activity in the south-western states, therefore, was bound to have a major impact on Denver.

There is, however, a paradox about the city of Denver. On the one hand, it has always been both the economic kernel of the state and its capital; by western standards, at least, it was an old major city and, as we shall see, it also practised a form of old-style party politics. On the other hand, much of Denver was always suburban in character – it was, and remains, a predominantly white and middle-class city. In 1980 only one resident in eight was black, and a much higher proportion of blacks were in white-collar jobs than in other cities. The poorer, western districts of Denver have become predominantly Hispanic, and there is a much smaller middle class in this ethnic group than in the black community. Yet this group too remained small – at least in comparison with other major cities; in 1980 fewer than one person in five in Denver was of Spanish origin. Despite some migration by whites to suburban counties, and a decline of about 5 per cent in its population in the 1970s,

Denver did not become a poor centre-city area rapidly being abandoned by its white residents. It was not an example of urban decay.

If New York was pre-eminent in the nation and Denver was the major city in the Rocky Mountain region, the East Bay area of California could make few claims to leadership. Nevertheless, although they had always been overshadowed by San Francisco across the Bay, the cities of the East Bay were an important economic unit. The principal county in which they are located, Alameda, had a population of 900,000 in 1960, making it of similar size to the Denver metropolitan area.[5] Gertrude Stein may have been correct in claiming that the problem with the county's major city, Oakland, was that 'when you get there, there isn't any THERE there', but in 1960 it was till the thirty-third largest city in the country. Adjacent to Oakland is the city of Berkeley, the site of the main campus of the University of California and in the years between the two world wars widely regarded as one of the most desirable cities in which to live.[6] In the immediate vicinity of this northern part of the county is a number of much smaller cities, Alameda City, Albany, Emoryville, and Piedmont. The very division of the area into these self-governing units helped to accentuate the backwater character of the East Bay; it was often forgotten that Alameda County had a much larger population than the city–county of San Francisco. Like San Francisco a port and commercial centre, Oakland had originally developed as the terminus of the trans-continental railway. It had been a large city before the Second World War, but in common with other California cities its growth was affected significantly by the war economy. Between 1940 and 1950 the populations of Oakland and Berkeley increased by 27 and 33 per cent respectively. However, because of the limited amount of land within the city boundaries, the population increase was even greater in the relatively undeveloped parts of the county. After 1950 Oakland's population started a slow, but continuous, decline, while Berkeley's population stabilized until the 1970s when it experienced a 9 per cent decline over that decade. From the 1950s the small towns in the southern part of Alameda County expanded, as did the new towns in the eastern part of the county from the 1960s. With the new businesses and new suburbs located elsewhere in the county, the relative importance of the Oakland-Berkeley area in county and state government was reduced. In 1940 77 per cent of the residents of the county lived in either Oakland or Berkeley, by 1960 only 52 per cent of them did so, and by 1980 a mere 40 per cent of the county lived in these two cities.

At the same time there was a transformation in the racial composition of the area. By 1950 15 per cent of Oakland's population was non-white; and this trend continued so that by 1980 only 38 per cent of the residents were white. Increasingly, the Oakland 'flatlands' came to be inhabited by blacks while the hills remained the preserve of middle-class whites. The geography of Berkeley is similar, but here the unusual circumstance of a large university campus dominating the city restricted black immigration, by putting working-class blacks into competition with students and others for the available housing. After a big surge in black immigration in the 1950s, when the non-white population of Berkeley increased from 15 to 26 per cent of the total, there was no increase in this population over the next twenty years.

There are two distinct, and contrasting, sides to the economic profile of Oakland. On the one hand, it displayed the usual characteristics of large cities abandoned by their former inhabitants. These included high unemployment, large numbers of people living below the poverty line, and a high crime rate. Although it was not quite as well known for these problems as, say, Detroit, they were still sufficiently associated with Oakland for the script writers of a nationally broadcast situation comedy to use them as the main theme of one episode.[7] On the other hand, the city's economy had been remarkably successful in many ways, especially in the 1970s. Because it adapted quickly to the container trade, while San Francisco with its much stronger labour unions did not, Oakland became the busiest port in the western United States. By 1980 it was one of the six largest ports in the world, providing over 23,000 jobs. In turn this stimulated building development in the downtown area with new office blocks, hotels, and shops. This 'affluence in the midst of squalor' is easily explained. A good network of roads made it possible for those living in both the East Bay and the San Francisco peninsular to commute to their jobs; many of those employed in Oakland, including city government employees as well as port employees, no longer lived in the city.

To conclude: like New York, Oakland was popularly regarded as a city experiencing 'urban blight' during the second half of the period of our study. While it is true that both exhibited many of the problems of inner city areas, it is important not to exaggerate the extent of their decay. Unlike smaller cities dependent on a few industries, these areas were major trading and commercial centres and they cannot be characterized as being in a continuous economic decline. They were losing population, and with that would go some of their political power *vis-à-vis* their neighbours, but the main components of their economic bases remained intact. Thus, in looking for an explanation of the decline of the political parties in these cities, we could not look simply at economic collapse for an explanation, for there was no such collapse. These were major cities forty years ago and they remained so in 1980. The socio-economic changes which affected the urban Democratic parties in America are far more complex than an abandonment of the inner cities.

## 2. Political and Electoral Institutions in the Three Areas

Any attempt to explain why a particular style of party politics emerged in a given city or state must take account of its political culture and history. Indeed, these are the central factors in explaining the rise, and persistence, of political machines in some cities but not in others. For example, Martin Shefter has argued that the much greater success of Progressive reformers in the western United States is related to the weakness, and the narrower base, of the parties there in the pre-reform era.[8] Patronage had been a vital part of western politics, but the parties had been more peripheral in the patronage system because the dominant economic interests were controlled by those outside the state; there remained many unmobilized potential voters in the west, and the Progressives were able to exploit this. Nevertheless, political institutions, which are the product of the historical experience, themselves come to influence the political style of an area; their influence can persist long after the particular political force which helped to create them has been weakened. In other words, political and electoral institutions provide 'boundaries' within which certain political practices can become established. These 'boundaries' become important in explaining transformations in electoral competition which occur either when party organizations are unable to campaign adequately for their candidates, or when new campaign resources become available to candidates. Of course, it is not just the institutions of the city or county which are important in determining local 'boundaries', but those of the state and federal union as well. To understand the development of particular parties, then, we need to examine the institutions of three levels of government. The constraints on party development imposed by a federal union employing the separation of powers are well known, so that here we need consider only the impact of particular city and state institutions.

In introducing the subject, we must draw attention to a point that is obvious, but sometimes overlooked: governmental structures at the state level in America are remarkably uniform, while those at the local level are extremely diverse. Every state practices the formal separation of powers, in imitation of the federal government. All but one of the states (Nebraska) copy the federal union in having two legislative chambers with comparable, if not always equal, powers. This has limited the variations to be found in executive-legislative relations and in party structure. Of course, the states *do* vary in respect of the powers of governors, the professional expertise available to legislators, and in other ways, and this provides for heterogeneity. Nevertheless, the government of any American state is much more like that of any other state (and the federal government) than it is like most other liberal-democratic governments.

In the governance of American cities there is no single model from which all others are derived. Most cities in America have either a mayor-council or a city-manager form of government.[9] There are so many variations on both of these forms that making general statements about the differences between them is difficult. The main difference is that in mayor-council systems executive responsibility lies with the mayor, or with the council, or is shared between them. That is, it rests with those who are popularly elected. In the city-manager system executive power rests with a manager appointed by the council to whom he is responsible. The council acts as a legislative body or, as it were, as a 'popular' check on a non-elected executive. Of the cities in our study, New York and Denver employ mayor-council systems while Oakland and Berkeley have a city-manager, although both also have elected officials called mayors. The city-manager was, of course, a device introduced in the Progressive era in the hope of reducing the power of parties in local government to, at most, a checking role.

Another Progressive device for limiting the power of parties was the non-partisan election. As a category for political scientists, 'non-partisan election' is one of the most misleading; all it means is that a candidate's party designation is omitted from the ballot paper. On this basis, no British general election before the 1970s could be considered partisan. Not surprisingly, studies have revealed that in non-partisan elections in America party activity varies. It ranges from being extensive, with party labels used regularly in campaigns and party organizations involved in the recruitment of candidates, to non-existent.[10] By itself the non-partisan ballot did not guarantee that parties were removed from local politics, but, undoubtedly, in many places it helped to reduce their role. Some cities, including Denver, adopted non-partisanship but retained the older, mayor-council, system of government; indeed, Denver retained a system which had one of the most powerful mayoralities in the country. Of the three other large cities in our study, New York has an arrangement which is common in its region (the partisan ballot), while Oakland and Berkeley, like all California cities, have non-partisan ballots.

We began this discussion by looking at cities, but we must now recognize that the relationship between city and county is different in the three areas. Local government is more fragmented in Oakland and Berkeley because there are two separate units of government, the cities themselves and Alameda County. *Ceteris paribus*, this division of functions between two elected bodies would make party involvement in local government less effective. But, of course, this is only one of several factors which help explain differences between levels of party activity in cities. Denver is the exact opposite of the East Bay – it is both a city and a county. However, like the East Bay and most American cities, one of the major services provided in Denver, schooling, is controlled by an independently elected school board. In most other respects, though, responsibilities and powers are centralized in Denver, in the office of mayor. In comparison with virtually all other major American cities, Denver stands out as a city which gives little checking power to the city council or any other body. In his ability to appoint policy-makers and to veto the legislation of the city council, Denver's mayor is more akin to the French President than the usual chief executive found in America. In contrast, although the mayors of Oakland and Berkeley were increasingly expected to be policy-making leaders of their councils, they continued to lack the executive power and the staff necessary to be more than legislative leaders.

The position of the New York mayor is different again. Three features contribute to this being the most important mayorality in America: New York is the largest city; New York has been the leading state in making provision for the 'needy'; and the city has run many services that elsewhere are financed and administered by independent boards or the state government.[11] However, if his is one of the most important executive positions in the country, the New York mayor is constrained in two ways which his counterpart in Denver is not, and these constraints affect his ability to act as a party leader as well as the leader of government. The very size of the city budget, and legal limits on the level of real estate taxation, mean that direct bargaining with the state government over the financing of city government is a normal part of the policy process. At the local level he is constrained by the Board of Estimate, a body which has no counterpart in Denver and which helps preserve the five constituent counties of New York City as units of local government.

At the risk of oversimplification, the Board of Estimate is best described as a kind of upper chamber in a system in which the lower chamber, the City Council, is relatively powerless. The Board has eight members, one of whom is the mayor while the others are the city's Comptroller, the President of the Council, and the presidents of the five boroughs. As an upper chamber, the Board of Estimate fits neither the American model of separated powers nor the parliamentary model in which the executive can be removed by the legislature. For our purpose, in examing the political framework in which the parties must operate, its most significant feature is that the constituencies of the five borough presidents are those of their own counties, whereas the constituency of the three other members is the city as a whole. This was one of the factors which prevented the development of a unified, city-wide political machine after consolidation in 1897, and it has similarly influenced more recent changes in the party organizations.

We began by considering local political structures, rather than state structures, because of their great variation in America. Nevertheless, if the general characteristics of state government are uniform in America, the organization of state politics in the fifty states is sufficiently different to increase diversity in local parties. Five factors are especially important. One is the extent to which the state legislature is 'professionalized'. Some legislatures meet for only a few weeks each year, have few research facilities for the members, and provide only a small renumeration for them; by 1980 others were providing conditions of service which were better than some national legislatures in the western world. This affects local parties because of its influence on both the structure of the political career 'ladder' and the resources available to legislators in becoming less dependent on the party's campaign resources.[12] The second factor also affects the career structure – this is the number of state legislators there are in each congressional district. The larger the state legislature, and the smaller the population of the state, the more potential rivals there are for an incumbent congressman; the more rivals there are, the smaller are likely to be their individual power bases, and hence the threat to the congressman is lessened. The fewer the state legislators in his district, the greater the incentive for the congressman to develop a power base independent of a party which might sometime turn against him. Thus, when it is possible for candidates to campaign independently of their party, we would expect that a party organization would come to have less influence when the state is large, and when the legislature is professional and small. The structure of other offices, contested locally, can have a similar impact on party organization.

A third factor is the type of electoral districts used in election contests, especially for the state legislature. Multi-member districts increase the potential for party control over the nomination process, and the larger the number of seats contested in a district, the greater that control can be. Single-member districts tend to minimize party influence because the party must retain a sufficiently strong organization in all districts if it is to control every nomination. Similarly, state law may regulate the type of primary election a major party can use in selecting its candidates. Party power can be maximized when some form of convention is used to determine which candidates are allowed to enter the primary ballot; it is likely to be at its minimum when any would-be candidate can gain easy access to the ballot. The final factor is also an aspect of state regulation of the parties. In states where the law requires parties to organize at the precinct level, and lays down how these lowest-level officials shall be selected, the party can retain an organizational structure even when it is having difficulty recruiting activists. When there is no legally-defined precinct organization, a party has less incentive to recruit activists in precincts in which recruitment is especially difficult. A precinct organization will not prevent the decay of a county party, but it might be expected to slow down that process.

Obviously, our three areas vary considerably with respect to these five factors. California and New York were among the first states to have 'professional' legislatures. In the 1960s each of them improved the conditions of service of members, turning what had been the best of the 'semi-professional' legislatures into well-paid, well-staffed, full-time bodies. Colorado has remained in a 'lower league' from the two larger states, but the conditions in its own legislature did improve significantly in the 1970s as did conditions in many of the legislatures of small- and medium-sized states. The ratio of congressmen to state legislators is also very different in Denver: this is about 1:25 at the beginning of our period and reduces to about 1:20; in New York the ratio has been about 1:5 and in the East Bay about 1:3. A third feature of the institutional structure of state politics in Colorado was also highly favourable to party influence: until the mid-1960s Denver elected its seventeen assemblymen and eight senators in county-wide multi-member districts. In both New York and the East Bay single-member districts were used.

Despite Progressive influences which, for example, led to the introduction of the initiative referendum, Colorado had political institutions which were at least as supportive of highly-organized parties as those to be found in New York. Unlike many western states, a 'framework', in which parties could play a major role in the electoral process, was preserved in Colorado. Yet, clearly, Tammany Hall would not have survived so long had there not been some favourable elements in the institutional environment of the state. When we turn to the last two factors, we can see that, indeed, there was a 'framework' supportive of party politics in New York City, as well as in Denver, while in the East Bay there were virtually no institutional supports for it.

Of the three states, California most clearly restricted party control in the nomination process. Any would-be candidate could file to enter a party primary, and to do so he was required only to pay a filing fee and to obtain the signature of a small number of supporters who were registered voters (between 20 and 100 depending on the office). Party voters then chose the candidate at a primary election. Until 1959 this was not even a genuinely partisan primary because, under the cross-filing primary law, a candidate could enter the primary of more than one party. If he won both the Democratic and Republican nominations, the candidate would then face, at most, minor party opposition at the general election. The effect of this provision was to aid incumbents, for they could exploit voter familiarity with their names to win the primary of the other party, thereby snuffing-out potentially dangerous opponents at an early stage.[13] For non-state-wide offices, the *formal* requirements for getting onto the ballot in New York were not that different from those in California: candidates had to obtain signatures to enter a party primary, but the number required was higher than in California. (For example 350 were needed by a would-be assemblyman and 5,000 by a would-be mayor.) This larger number, together with the partisan judicial system in which election disputes were heard, created a barrier to entry onto the ballot which the parties could control. For state-wide offices in New York there were no primaries after 1922, candidates being chosen by party conventions. This provision changed in 1967, and contested primaries were to become a feature of nominations for major offices after that. In Colorado there was always a formal role for the party in nominating candidates at all levels of partisan office. Parties have employed pre-primary nominating assemblies in which candidates have to obtain 20 per cent of the vote in order to get onto the primary ballot.[14] This meant that, in theory, there could be as many as five candidates contesting a primary; in practice, there were usually two, and sometimes three, candidates designated for the primary. Party influence was possible because delegates to county assemblies were chosen at neighbourhood caucus meetings, and these were meetings organized by the members of the county party.

Finally, in explaining how the institutional framework in California was not conducive to strong parties, we must consider the absence of legally-defined precinct organizations. In California there have been no formal party organizations below the county committees, so that Californian parties remained more like the pure kind of caucus-cadre party than those in New York. In Denver, as in New York, there have been precinct organizations, though, in discussing county parties, care must be exercised in distinguishing the lines of authority recognized in law from the sources of *de facto* authority. In Denver precinct committee members have been elected at the same time as voters nominate candidates for partisan public offices; however, when vacancies occurred before an election, they could be filled temporarily by appointment of the county party chairman – an obvious source of potential power. The committee members obtained their designation to seek the nomination from the same precinct caucuses at which delegates to the county (and district) assembly were elected. These have been the only legally sanctioned elections in the party organization, but the body which they help to form, the County Central Committee, has been too large to be effective; its potential membership is over 1,000. The Democratic party overcame this problem by having the committee members select both party officers (including the chairman) and a captain (and co-captain) for their district of the city. The officers, captains and co-captains have then formed an Executive Committee which, with between 70 and 100 members, has been the size of body which could meet regularly.[15] Since this was the only effective decision-making body, the politically ambitious captains had an incentive to control nominations for precinct committee members in their districts: their political future depended on the votes of these members. While they could organize locally to bring in to the party the people they wanted, they also had to work with a county chairman whose power of appointment made him more than *primus inter pares*. However, given that turnover among committee members was always likely to be greater than among (the more committed) captains, often the latter could create the kind of organization they wanted, at least in the short term, through bargaining with the chairman.

Of course, the model for party organization in Colorado was taken from that found in many eastern states. Nevertheless, there have been considerable variations in the model, and there were significant differences between the party structure in Denver and that in New York. In New York candidates for the County Committee had to obtain signed petitions to get onto the primary ballot; the Assembly District Leaders (the equivalent of Denver's captains) were directly elected in primaries in Manhattan, Brooklyn, and Queens, though not in the Bronx and Staten Island; the Assembly District leaders appointed Election District captains, an intermediate position not needed in the smaller city; and, finally, the Assembly District leader owed his position to his leadership of the dominant political club in his district. These last two points are critical in understanding the New York party. The power to appoint key officials under him gave the Assembly District leader partial control of the effectiveness of election activities in his district. In a system of patronage politics, it made him the most important actor below the level of county leader; in an era of challenges to patronage politics, it helped to make him immune from reformers whose organizational strength lay outside his district. He became a district leader because of the strength of his club – it was the organizing ability of the club which permitted its leader to assume a position in the legally-defined party. Challenges to him came from other clubs. This element of a traditional caucus-cadre organization operating behind the legally-defined organization was absent in Denver, where party institutional structures did not protect the old-guard so well against insurgents.

We must reiterate that, while the institutional framework does not determine what form party politics will take in a particular place, it does influence it, and the parties seek to create a framework in which they can most easily operate. This is no great revelation, but it is sometimes overlooked when examining the social causes of party decline. We have seen that in New York the structure of local and state political institutions was of a kind supportive of party activity. In Denver the institutions at the state level were even more favourable to parties than they were in New York, but the non-partisan nature of local government there provided a potential source of weakness for the decentralized parties of the caucus-cadre type. Finally, in the East Bay the institutional framework was about as weak as it could be for a continuous and extensive electoral role for parties. As we shall see, these different frameworks have been important influences on the ways in which the parties have responded to internal decay and new competitors.

## Notes

[1] On the process of consolidation, and for a more detailed account of New York political history, see Wallace S. Sayre and Herbert Kaufman, *Governing New York City* (New York: Russell Sage Foundation, 1960), ch. 1.

[2] In fact, the Bronx did not become a separate county until 1914. The county names for these boroughs differ in three cases – New York (Manhattan), Brooklyn (Kings), and Staten Island (Richmond).

[3] Of the border states, three (Oklahoma, West Virginia, and Kentucky) had increasing populations while two (Maryland and Missouri) had declining ones.

[4] In spite of new building development sprawling eastwards across Long Island, the population of Nassau and Suffolk counties increased by only 2 per cent in the 1970s.

[5] The East Bay is even larger than this, because the industrial area of Richmond is in neighbouring Contra Costa county.

[6] Of this period T. J. Kent, Jr., has claimed, 'In the 1920's and 1930's Berkeley was a dream city. In more ways than anyone was conscious of at the time, it had become one of the world's best examples of the kind of ideal community suggested in 1898 . . . by Ebenezer Howard . . .', 'Berkeley's First Liberal Democratic Regime, 1961–1970: The Postwar Awakening of Berkeley's Liberal Conscience', in Harriet Nathan and Stanley Scott (eds.), *Experiment and Change in Berkeley* (Berkeley: Institute of Governmental Studies, University of California, 1978), p. 73.

[7] *New York Times*, 27 Jan. 1981. The point of the episode was to contrast 'civilized' San Francisco with the 'jungle' of Oakland.

[8] Martin Sheifter, 'Regional Receptivity to Reform: The Legacy of the Progressive Era', *Political Science Quarterly*, 98 (1983), 459–83.

[9] About 5 per cent of cities use the commission system, in which separately elected commissioners have independent executive responsibilities; some smaller municipalities use devices such as town meetings.

[10] Charles R. Adrian, 'A Typology for Nonpartisan Elections', *Western Political Quarterly*, 12 (1959), 449–58.

[11] Charles R. Morris, *The Cost of Good Intentions* (New York: Norton, 1980), p. 127. These services include education, higher education, and the welfare system.

[12] The idea of career structures is formalized in Schlesinger, *Ambition and Politics*.

[13] In fact, the abolition of cross-filing came in two stages. From 1954 onwards candidates had to reveal their party affiliation on the primary ballot, and it was finally abolished in 1959.

[14] There was a provision for direct access to the ballot through a petition procedure, but until 1980 this had always proved difficult for most would-be

insurgents; in 1980 the eventual Republican nominee for the US Senate gained access by this route. There were two difficulties – the number of signatures required was large (500 in the case of the state assembly), and it was expensive (signators had to state on oath before a notary public that they would vote for the candidate).

[15] The size of the Executive Committee increased from about 70 in the 1950s (of whom, nearly half – the women co-captains – were often 'second class' members) to 100 in the 1970s.

# 3
# 1940-1960: Indian Summer for the Parties?

In this chapter we aim to show what happened to the parties before competition from the new campaign technologies became effective. We look at each of the three areas in turn, but devote less attention to New York because this case is much better known, and is well documented.[1] The main thrust of our argument differs from a popular contemporary view, for we claim that in the 1940s and 1950s the parties were not becoming so weak that complete collapse in the 1960s was inevitable. Far from being in continual decline since the height of the New Deal, in some respects the parties actually had a brief revival about the middle of the century. Nevertheless, as we shall see, this revival disguised an underlying weakness in their condition which became critical when they faced competition from the individual candidate with his independent electoral resources.

Today the conventional view of the 1940s and 1950s is of a continuation of longer-term party decay. While writers in the early 1960s had generally seen those years as part of an era of relative stability for the parties, following the assault on them by reformers in the period 1890–1920, more recent writers emphasize a rather different feature. A typical view is that of Morris P. Fiorina: 'The New Deal period temporarily arrested the deterioration of the party organizations, at least on the Democratic side. Unified party control under a "political" president provided favorable conditions for the state and local organizations. But following the heyday of the New Deal . . . the decline continued.'[2] In this view, the New Deal was merely a hiccough in a trend which started only thirty or forty years after the growth of relatively centralized, patronage-based, urban organizations. Now it is true that the New Deal provided some sustenance to state and local parties, that many party machines lost power in the 1940s, that the electorate learnt to split its tickets in the Eisenhower era, and that in the 1950s party organizations faced the emergence of middle-class, amateur activist groups. Yet there is another aspect of the period 1940–60. Where the old guard were willing to compromise with the new elements, the parties were actually strengthened because many of the new activists were party-oriented; at the national level the party organizations retained their power of veto over presidential nominations, as Estes Kefauver discovered; and, at the local level of politics, the power of individual candidates in the nomination process, *vis-à-vis* party organizations, changed very little in the twenty years after 1940. Indeed, at levels of politics below that of the national level, there is some justification in describing it as the last era of 'old-style' party politics in America. To understand this point, it is useful to turn to that master interpreter of American politics, V. O. Key.

Before doing so, however, it is worth explaining why the conventional view can be misleading. First, it may incline us to accept more readily what we called in Chapter 1 the 'myth of party decline' (the argument that without decline the campaign consultants could not have taken over), before considering whether there are basic features of American parties which made them susceptible to take-over by the consultants. Secondly, if what happened to the parties in the 1960s is seen simply as part of a process of long-term decline, we are less likely to focus on the separate influences that a number of factors had on party collapse. Several of them – changes in the ethnic composition of America's cities, the decline of labour unionism, and institutional reforms such as reapportionment – developed after the New Deal, and are only indirectly linked to the sources of party decline in the years between the 1890s and 1920s. The disentangling of these separate influences is central to a full understanding of party decline. Finally, by focusing on very long-term trends, we may ignore short-term reversals in these trends. While it is now widely accepted that the New Deal partly reversed earlier party decline, it is less widely acknowledged that some developments in the 1950s were supportive of party-centred politics. It is these developments that we examine in this chapter.

In *American State Politics* Key noted that, in states which had genuine two-party competition, control by the same party of both governorship and legislature occurred on only about half the possible occassions. He argued that the failure of parties to control both branches of government at the same time could be explained by two institutional factors: non-coinciding terms of office and malapportionment. In his examination of state elections from 1930 to 1950, he concluded that:

> Only infrequently does the electorate deliberately choose to place the executive and legislature in the hands of opposing parties, despite the prevailing impression to the contrary. By sifting through the records of

our 32 states over a period of 22 years, only a half dozen or so instances can be isolated in which such was apparently the case.[3]

According to Key, if the institutional impediments were removed, divided control would become rare. It remained common, however, in the two-party states throughout the 1950s, and was absent only when a 'national landslide' overcame the influence of the institutional factors. Consider the sixteen states which Jewell and Olson identify as competitive two-party states in the period 1946–58.[4] In state elections between 1954 and 1958 there were twenty-nine instances of simultaneous elections for a governorship and an entire lower chamber of the legislature; in eighteen cases the same party won the governorship and control of the lower chamber, in ten cases there was divided control, while in the remaining case the two parties had the same number of seats in the legislature. Only in the 'landslide' year of 1958 do we find no instance of divided control; it took a national surge for one party to overcome the effects of gerrymandering.

Although non-coinciding terms of office remained, malapportionment (though not other forms of gerrymandering) was removed by the Supreme Court's decisions from 1962 onwards. It might have been expected that this would reduce the incidence of divided control, so it is instructive to compare a post-reapportionment period with the period 1954–8. We have selected the period 1970–4 which, like the earlier one, includes a major landslide year (1974). There are twenty-three states with partisan legislatures which Jewell and Olson identify as competitive for the period 1960–76, and these generate thirty-eight instances of elections for governorships and entire lower chambers between 1970 and 1974.[5] The instances of control by one party outnumber those of divided control by 25:13, which means that 66 per cent of all cases involved single party control, as against 64 per cent for the earlier period. Why, then, did reapportionment not have the impact that Key suggested it would? A clue to this puzzle lies in the results of 1974. Unlike 1958, this 'landslide' year produced four instances (out of fourteen) in which the beneficiaries of the tide failed to win both governorship and lower chamber.[6] What had changed since the 1950s was that voters had come to choose different parties for a governorship and state legislaure much more frequently. To see how common a phenomenon it became, compared with the years observed by Key, it is useful to turn to slightly different data, in order to make certain we are excluding states where one of the parties was competitive in gubernatorial elections, but was not really competitive in state legislature elections. That is, we adopt a more restrictive definition than the one devised by Ranney, which Jewell and Olson use.[7] We employ the Ranney criteria, but then require that a state is to be rated as competitive only if both parties had control of governorship, upper chamber, and lower chamber, for at least four years each in the period 1965–78. (1965 can be taken as the beginning of the reapportionment era.) This reduces to eleven the number of competitive states (see Table 3.1). Of the thirty-six instances between 1965 and 1976 involving both a gubernatorial election and an election for an entire lower chamber, there were eleven occasions when control was split between the parties. These were not cases of institutional factors frustrating a popular will, but, in Key's terms, of an electorate choosing to divide control of government. (Of course, only a minority of voters may have been choosing this, but their voting power was crucial.) From being a rarity, it was occurring in about 30 per cent of elections by the 1970s, while the continued impact of staggered elections for state senates meant that on only about 55 per cent of all occasions was there single-party government in the most competitive states. The point of drawing attention to this transformation is to suggest that, in a real sense, the 1950s could be said to have been a decade of party politics in America, albeit one in which institutional factors continued to frustrate party control of state government. Yet it was to prove to be the last decade of that dominance for the parties.

If party voting remained high, at state levels of politics, in the 1950s, there were other reasons for believing that the parties were not decaying and might, possibly, be experiencing a regeneration. Party competition in many non-southern states was increasing, party identification in the electorate was high, the parties retained about as much control over the nomination process as they had in the 1920s and 1930s, and they were attracting some elements of the middle classes into party organizations. But, of course, this view of party politics in the 1950s must be set against one which is emphasized today – the withering of party activity. Even by the early 1960s there was considerable evidence that much of the party workforce was relatively inactive. In the old machine city of St. Louis Salisbury found that 27 per cent of the organization activists did not perform any significant electoral tasks, while in Detroit Eldersveld had found that 10 per cent of Democratic precinct leaders did not perform *any* political activities. Indeed, in Detroit only 17 per cent of the precinct leaders performed all three of the campaign tasks which Eldersveld identified as critical.[8] Whether they had ever been as efficient as they were popularly thought to be may be doubted, but

**Table 3.1**: Control of Governorships and State Legislatures by Party, following Gubernatorial elections in two-party competitive states, 1965–1976

| | 1966 | | | 1968 | | | 1970 | | | 1972 | | | 1974 | | | 1976 | | |
|---|---|---|---|---|---|---|---|---|---|---|---|---|---|---|---|---|---|---|
| | G | S | H | G | S | H | G | S | H | G | S | H | G | S | H | G | S | H |
| Delaware | | | | R | R | R | | | | D | R | R | | | | R | D | D |
| Pennsylvania | R | R | R | | | | D | D | D | | | | D | R | D | | | |
| Arizona | R | R | R | R | R | R | R | R | R | | | | D | R | D | | | |
| Iowa | D | R | D | R | R | R | R | R | R | R | R | R | D | D | D | | | |
| Nevada | R | D | D | | | | D | D | R | | | | D | D | D | | | |
| New Jersey* | D | D | D | | | | R | R | R | | | | D | D | D | | | |
| Utah | | | | D | R | R | | | | D | R | R | | | | D | D | R |
| Illinois | | | | R | R | R | | | | D | R | R | | | | R | D | D |
| Wisconsin | R | R | R | R | R | R | D | R | D | | | | D | D | D | | | |
| Indiana | | | | R | R | R | | | | R | R | R | | | | R | D | R |
| Ohio | R | R | R | | | | D | R | R | | | | R | D | D | | | |

*Abbreviations:*
D Democrat,
R Republican,
G Governorship,
S Legislative Upper Chamber,
H Legislative Lower Chamber,
* New Jersey elections held in 1965, 1969, 1973

certainly by the late 1950s most party organizations were not the 'well-oiled machines' which might have mounted some resistance to the challenges they were to face. Moreover, there were some parties in which internal conflict among activists in the 1950s weakened the party organization and reduced party loyalty to it. Perhaps the best example of this in the country was the Democratic party in New York, a case which a proponent of the conventional view of the 1940s and 1950s might well point to when elucidating the theme of party decay. In this chapter we consider this case first before turning, in the two subsequent sections, to examine cases where party decay was not apparent and where the Democratic parties appeared to be enjoying an 'Indian Summer'.

## 1. New York City: Continuing Party Decay

While political machines in many cities were collapsing soon after the end of the Second World War, the Democratic machine in New York City won back the mayorality which it had lost twelve years earlier to Fiorello La Guardia. Yet this victory did not lead to the revival of the city machine for, as in cities where formal control was being ceded, the Democratic organization in New York faced serious problems. Unlike Chicago, where a decade later Richard Daley was able to rebuild a decaying machine, attempts to do this in New York merely created new difficulties for the party. The very structure of New York City government had made party management a formidable task ever since the city's consolidation. In a real sense, there was no New York party, but five borough parties operating in a federal-type structure, with two of them (Manhattan and Brooklyn) vying for leadership. For all the popular references to Tammany Hall's power, the Manhattan party controlled the mayoral office for only about fourteen of the sixty years following consolidation.[9] When it had not lost control to the Brooklyn party, Tammany Hall had been defeated by one of a succession of reform movements which emerged as challengers to the machine style. In this respect, those who stress that American party decline has been a very long-term process have an important point to make, for reform did more than keep machine politicians out of office. It changed the nature of local government institutions. Members of the middle-class professions, who were leaders in the reform movement, restructured local government to provide fellow professionals with opportunities to practise their skills.[10] As Shefter has noted:

Because reformers in New York and throughout much of the Northeast have enjoyed only episodic success in the electoral arena, they have concentrated their efforts on creating public agencies insulated from the influence of electoral politics – professionalized police and welfare departments, independent authorities, financial control boards, and the like.[11]

The result was a reduction in the governmental domain which the machine could control for purposes of patronage, and these changes were not reversible.[12]

To point this out, however, is not to suggest that the Democratic organizations in New York City were to experience electoral failure, for they did not. Nevertheless, they were perhaps most effective when harnessed to an imaginative and powerful Democratic incumbent in the governor's mansion in Albany. This was the situation during Alfred Smith's tenure in that office (1919-21 and 1923-9). On the one hand it produced a major shift towards state responsibility for social welfare. Of this, Gifford has said: 'Other states such as Ohio, Illinois, and Pennsylvania also boasted large and ethnically heterogeneous populations. They had their counterparts to Tammany Hall, but none of their political machines supported sweeping reform measures comparable to those backed by the Smith-Murphy, Tammany-based New York Democratic party of the 1920s.'[13] On the other hand, this partnership generated patronage for a party that was denied much of the usual access to federal patronage during the presidencies of Wilson and Franklin Roosevelt, because it backed their Democratic opponents during their successful bids for the party's nomination.

The opposition of Wilson and Roosevelt caused further problems for the party, for each backed successful insurgent candidates against the Tammany machine the year after his own election. In the case of La Guardia, this was to deny the Democrats City Hall patronage from 1933 to 1945. In those years they had to sustain themselves at the local level from patronage provided by the Surrogates Courts.[14] Yet the Democratic parties in all the boroughs survived the La Guardia years. It was, however, a survival of old-fashioned parties that had been little affected by the New Deal revolution. Despite its partnership with Al Smith in the 1920s, the party organization itself was not opened up to the new elements of the Roosevelt coalition. Individual 'New Dealers' and liberal labour unions came to be organized in New York through their own political party (the Liberal party) rather than, as in other cities, being embraced within the Democratic party. This meant that a party which was later to face the challenge of the 'amateur' Democrats was one with no tradition of accommodating new elements. Coupled with a tradition of borough-based factional disputes, it was to prove a recipe for disaster.

By 1950 the party had again lost the mayoral office. In some respects, this followed the usual pattern of failure – a 'substantial number of Democratic District leaders' defected to an insurgent candidate because of dissatisfaction with the party leadership.[15] This was enough to tip the electoral balance for the dissident Democrat, Vincent Impellitteri, in the special election that year. It was 'usual' in that it was regulars (not reformers, for there were none in the party organizations then) who were involved, and internally it produced shifts of power between clubs in certain districts. This time the loss of power in City Hall lasted only three years, and during this period the Manhattan party was regrouping under Carmine De Sapio who had been chosen county leader in 1949. At first, De Sapio's method for reuniting the party included procedural concessions to the nascent reform movement, such as the direct election of Assembly District leaders; later in the 1950s he was to return to a more traditional style of 'freezing out' dissidents.[16] Neither of these styles stopped the growth of the new reform movement – a movement which only later was to have the benefit of a number of 'blue chip' party people associated with it. In the early 1950s, though, the central problem for the borough leaders was that opposition to party leadership was coming from closer to the organization itself than it had done in the past. Impellitteri was a former President of the City Council, not an 'outside', anti-Tammany Democrat, as John Mitchel had been in 1913. Though it may not have appeared so at the time, the 1950 election can be seen as the first stage in the separation of the candidate from the Democratic party. The next stage of this process was when Impellitteri's successor Robert Wagner broke with De Sapio and the party leadership, and contested the 1961 mayoral election as a candidate running against the bosses. Among the groups that could take credit for his victory that year was the reform club movement which by then formed a substantial minority on the Executive Committee of the Manhattan county party. However, at that stage the movement was a Manhattan phenomenon – in no other borough had they gained control of any Assembly Districts. This point is important, because it shows how the separation of candidate and party was developing in New York while the reform group was still a minority movement found in one borough.

Although the amateur Democrats were the beneficiaries of major demographic changes in Manhattan, the reform clubs faced a serious obstacle before 1958 – they were politically isolated. Potential allies either operated in other arenas, such as the Liberal party, or, as in the case of blacks, saw little future in co-operation with the reformers. The event which was to give this small movement a more influential part in New York politics was the controversial state party convention of 1958.[17] Whether it was a case of De Sapio pushing too strongly for the nomination of his candidate for the US Senate, or the interaction of divided party leaders and a dithering Governor Harriman, need not be examined here. What is undeniable is that in his re-election year Harriman had as a senatorial candidate someone he did not want, and this was widely seen as the assertion of 'boss power'. The result was that in a year of landslides for Democrats throughout the country, the Democratic gubernatorial and senatorial candidates in New York both lost. In turn, this defeat was to push 'blue chip' leaders such as Harriman and Eleanor Roosevelt closer to the reform movement, and it was to keep state patronage away from the Democratic organizations for sixteen years. The amateur Democrats were immediate beneficiaries of the débâcle, as Wilson has shown, 1960 and 1961 were years of great progress for the movement.[18] However, it must be emphasized that progress had been made possible not just, as Wilson stresses, by the decline of working-class neighbourhoods, but by the division among its opponents in the city. Nevertheless, as we shall see, in the longer term it was not the reform movement which would benefit from the persistence of conflict in the New York parties, but the individual candidates.

If in New York the 1940s and 1950s were decades of conflict during which the Democratic parties undermined their own position, this experience does not seem to have been a common one in the rest of the country. Despite a disastrous election in 1946, and the defeat of their presidential candidate in 1952 and 1956, dissension within the non-southern parties was generally contained. Stevenson's candidacy may have helped to mobilize new activists, but his subsequent defeats did not produce regular-reform splits in Democratic parties. Nor, in 1960, did resentment of Kennedy's nomination by Stevenson's amateur supporters lead to anti-party activism on their part. At the local and state levels too we find little evidence of the widespread party chaos of the kind found in New York. There were disputes between amateur and professional activists in several places but these did not produce the civil war of the New York party. As we see in the Denver study, co-existence between amateurs and professionals was possible; where there were no professionals, as in the East Bay, amateur activism could actually rejuvenate the party. It is to this latter case that we now turn.

## 2. The East Bay: Party Revival

Between the end of the Second World War and the end of the 1950s California politics changed radically. In particular, two of the main legacies it inherited from the Progressive era were removed. One was the chameleon-like character of its Republican party, a party which embraced both the wealthy conservatives based in southern California and a more reformist group. Being all things to all people, it remained the dominant party in the state until the 1958 elections. Earl Warren, the state's Governor at mid-century, was the personification of its ambiguous nature. Acceptable to the conservative power-brokers of Los Angeles, but regarded as a sufficiently orthodox Republican to be appointed Chief Justice of the Supreme Court by Eisenhower in 1953, Warren went on to become the touchstone of liberalism during his sixteen years on the Court. At the end of the 1950s there was to be a remarkable shift in Republican fortunes. In the twenty-eight years between 1933 and 1960 the party had controlled the governorship for twenty-two years, the state Senate for twenty-four years, and the state Assembly for twenty years; between 1960 and 1980 it controlled the governorship for eight years, the state Senate for two years, and never had control of the state Assembly.[19] In this later period state politics was characterized by a conservative Republican party confronting a moderate-to-liberal Democratic party. The second legacy, the cross filing primary, was partly responsible for a delay in the realignment of party strength in state offices; it provided an electoral advantage to Republican incumbents in the late 1940s and early-to-mid-1950s. However, one product of Progressivism could never be removed – weak party organization. A variety of devices, including the direct primary with easy access to the ballot, the referendum, and non-partisan elections, prevented the growth of local party organizations of the kind found in the east, even after the quasi-colonial period of its economic history had passed. Like Washington and Oregon, and many other western states, California preserved a relatively pure form of the caucus-cadre party. Small local groups, in some cases backed by out-of-state interests, supporting individual candidates were the dominant actors in elections; it is not surprising that, when non-party political consultants first entered American electoral politics in 1933, it was in a campaign in California.[20]

In the early 1940s the East Bay was a Republican area. It sent Republican legislators to Sacramento and Washington, and both Oakland and Berkeley were governed by representatives from their business communities. The latter could run for office without introducing party labels into the campaigns, while the former's short-cut to re-election was to seek the nomination of both major parties. Buttressing the status quo was the assistance provided by the *Oakland Tribune*, the East Bay daily newspaper with the largest circulation. The first catalysts which were to assist in the breaking of this electoral monopoly appeared in the mid and late 1940s. One was the labour movement, the growth of which in the 1930s enabled it to become a major political organizer for the Democrats in those respects where party organization was deficient. In 1944 union support enabled George Miller to defeat an incumbent in a congressional district which included part of Oakland and all of southern and eastern Alameda county; Miller was to serve in Congress until he was defeated in a Democratic primary in 1972. Labour unions were also of prime importance in the 1947 Oakland city elections, in which the Republican establishment was defeated. Their limitation as a partner in the 1940s and 1950s was that their contribution to campaigns usually varied directly with the perceived threat to union interests.[21] The zenith of their electoral activity, and of union involvement statewide, was in the 1958 elections when there was a 'right to work' referendum on the ballot. Their effort in defeating this was linked to support for Democratic candidates, and they were instrumental in the election of many Democrats, including the new Governor, Pat Brown. In the East Bay inter-union co-operation was facilitated by the siting of many union headquarters in the same building. While the resulting unity was an important political resource, it was supplemented by the unions' ability to recruit campaign workers, by their weekly newspaper which was distributed to all union households, and by their financial contributions to candidates.

The 1947 Oakland council elections had provoked union involvement because of a bitter general strike in the city the previous year. Their attempt to replace the city's political leadership led them into coalition with a new political force: returning war veterans. A curious feature of the popular view of American elections just after the end of the war is that it focuses on just two events: the 1946 congressional landslide by the Republicans and the 1948 'upset' victory of Harry

Truman. In fact, in the three or four years after the end of the war there were a number of elections in which established politicians or parties lost power to candidates who were backed by informal groups of war veterans. The Oakland experience was not unusual. A group of young ex-servicemen held a series of private lunch-time meetings and decided to run candidates in the city elections. About three of them expressed an interest in being candidates and an organization, the Veterans Committee for Good Government, was formed around them. One candidate Joseph Smith, had acted as a lawyer for union members in workmen's compensation cases, and he was approached by the unions to be a member of a slate of candidates which they would back.[22] At this point the other veterans withdrew their candidacies in favour of Smith and the rest of the slate, who then campaigned under the title of the Oakland Voters League. Although leading Democrats were active in the campaign, and in the accounts of some participants helped to facilitate the merger between the veterans and the unions, the party label was not used in the election; it was believed that this would be counter-productive in a predominantly Republican city.[23]

This last point draws attention to an obvious difference between Californian and East Coast politics at that time. Although the candidates were running as a slate to remove covert Republican control of the city, theirs was not a party-centred campaign, nor did the slate members devote themselves to campaigning for the slate as a whole. Each of the candidates distributed his own campaign literature as well as the slate literature; despite the fact that the unions supplied most of the campaign funds (even Smith got 90 per cent of his funds from them) they did not demand, or even want, a fully integrated campaign. Yet, by California standards, this was a highly co-ordinated team effort, and it proved successful in electing Smith and three others. With the help of the 'swing' vote of one incumbent councillor who was not up for re-election in 1947, Smith was subsequently chosen as mayor. However, without a stable working majority on the council, the considerable incompetence displayed by some members once in office, and the outright opposition of the *Oakland Tribune*, meant that the group had lost power again by 1949. In 1951 those members of the former slate who sought re-election did so individually. Thereafter, succession to the city council was determined largely by appointment – appointees to the vacant seats using the advantages of incumbency at the subsequent general election.[24] (Republican-inclined council members who wished to leave the council would time their departure to provide an electoral advantage for their successors.) Nevertheless, if its long-term effects on Oakland city government were small, the slate effort of 1947 was instrumental in stimulating political activity among Democrats in the East Bay.

A third element in the Democratic upsurge was the end of the a-political, 'dream town' era in Berkeley. There were three main causes of this change in the city's character.[25] The first was the huge increase in the city's population in the 1940s, an increase mainly in the poorer and non-white groups, so that by 1950 Berkeley was not simply a middle-class university town. Then in 1947 a scandal involving the mayor led to the election of a reformist clergyman to replace the miscreant, though the direct impact of this blow to Republican rule was even less than the 1947 elections in Oakland. Finally, in 1949 a loyalty oath controversy began in the University of California, the effect of which was to encourage political activism in what had previously been a quiescent faculty and student population. Civic disquiet was to lead to a remarkable growth in political organization from 1952 onwards, but its success ultimately depended on the continuing increase in the East Bay's black population.

By 1948 the rise in the black population in the East Bay's 'flatlands' was such that a black Democratic candidate could run sucessfully for a state legislature district. The labour unions had supported another candidate in the primary election, but his narrow defeat by Byron Rumford led them to switch their support to Rumford at the general election. Rumford was to remain in office until 1966. In a state with weak party organizations the capture of a public office was crucial for the development of a purely political organization in the black community. In its absence black ministers had a virtual monopoly as political leaders, and it was their initiative which had led to a black candidate running in the 1948 election. They had arranged a meeting at which the would-be candidates would be introduced, and after which a vote was taken among the ministers as to who would be chosen. It was not until after the election, in 1949, that a secular black organization, the East Bay Democratic Club, was formed. Affiliated to the Democratic party, this club of about 200 members was of central importance in the following decade. Several of its members were later to hold major judicial and administrative offices. In co-ordinating with the white organizations, its key member was D. G. Gibson. However, Gibson's role is the subject of some controversy. By some writers Gibson has been portrayed as the 'father' of East Bay black politics.[26] Indeed, it is true that he was a political activist from the 1930s and that he was, perhaps, the person most responsible for the founding of the East Bay Democratic Club. Yet whatever Gibson's campaign skills, Rumford would not have been elected in 1948 if it had not been for the leadership exercised by the church ministers. Moreover, two senior Democrats have argued that Gibson's power base was his close association with Rumford; they claimed that it was only after the latter's election that Gibson could become a political entrepreneur.[27]

Nevertheless, irrespective of his role before Rumford's election, there can be little doubt that Gibson was the vital link with white Democrats in the 1950s. During that period he was able at least partially to bridge a number of significant differences in political style and values between white and black politics. This link gave white candidates access to a loyal Democratic electorate. In some ways they faced a quasi-market view of politics held by the leading blacks – blacks saw white candidates as needing the votes they could deliver, and in return whites had to pay for them. The problem was that what the blacks most wanted, votes for their own candidates, could not be delivered by white activists; defection from black candidates remained high in the Berkeley hills. Leading white Democrats also believed that there was an imbalance in their partnership with blacks – but of a rather different kind. Black activism in political organizations was low and, while the black electorate was loyal its turnout was also low. Whites often put more effort into electoral organization, and then found that their candidates had to contribute money to the black churches in which they were not allowed to speak. This type of tension was much less serious than that which was to develop in the 1960s, but overt conflict was avoided only because of the co-ordinating role played by D. G. Gibson.

The 1950s was the era of change in party strength in the East Bay. By the end of the decade there was a flourishing Democratic organization; a Democrat had won the second congressional seat in Alameda County; and major advances had been made in Democratic representation on Berkeley city council, so that by 1961 Democrats took control of the council. As every student of American politics knows from James Q. Wilson's research, this period saw the growth of the California Democratic Council (CDC) club movement – a movement of mainly young, middle-class, liberal residents of large cities. Wilson argued that, except in their orientation to state, rather than city, politics, the clubs he studied in Los Angeles were similar to reform clubs in Chicago, New York and elsewhere.[28] Although his own study was concerned with the largest cities, he argued that there had to be a certain kind of 'cosmopolitan' middle-class population for an amateur movement to develop. In the East Bay there was such a population, though most of these people were resident in Berkeley and not in Oakland. Consequently, direct participation by this group in Oakland elections was usually confined to contests for offices with electoral boundaries which straddled the two cities – that is, to congressional and state legislature elections. Even so, there was some 'spill over' from the activism in Berkeley, for use was made of the county party as a mechanism for introducing partisan politics into Oakland city elections. For example, Eugene Lee reported that in 1959 the Central Committee of the Alameda County party endorsed a slate of Democratic candidates in the city council elections and sent out campaign literature for that slate. He also noted that this was 'the first occasion in recent years that a central committee had taken an official stand in school elections' in California.[29] At first glance, then, it might seem as if the East Bay experience was very similar to that in southern California. Nevertheless, there is one crucial respect in which the East Bay does not fit Wilson's model of California politics in the 1950s. In considering this in the present chapter we can lay the foundation for an explanation of why Democratic club politics disappeared in California.

Wilson's picture of the liberal activist movement in California was of a club-centred movement which, in southern California, derived its vitality from opposition to it at the County Committee level. He also saw it as geared to state, and not local, politics: in Los Angeles 'the political scoreboard is tallied on a state, rather than a city, basis', because 'unlike New York, it is impossible to examine the city apart from the state, for in California, partisan politics has had no relation to urban government.'[30] Whatever the reasons for the absence of interest in local politics in Los Angeles, the suggestion that it can be explained by the non-partisan nature of the ballot is implausible. Non-partisan ballots impeded partisan politics but were not a sufficient cause for its absence. Unlike Los Angeles, we find in the Berkeley area in the 1950s an amateur movement which was both state and locally oriented; this was reflected in the complex structure of organizations which existed there. It is this local orientation which helps to explain why a major club movement could develop in the East Bay, despite the fact, which Wilson recognized, that in northern California the party's County Committees were much weaker than in the south.[31] He was right to believe that political organizations, such as clubs, gain vitality from opposition to them. But where a movement has more than one focus of attention, it may sustain itself through opposition from one source even when another potential source is generating little conflict. This was why amateur Democrats thrived in the East Bay. To understand how they differed from the Los Angeles Democrats, it is useful to compare the organizational structure in the two areas.

In Los Angeles there were clubs linked to each other through the 'umbrella' organization, the CDC. In the East Bay there were three other types of organization, which were connected to the club movement and the CDC, but which had specialized functions. While they would not have developed but for the growth of a club movement, these organizations were distinct entities. Before describing them, it is perhaps useful to summarize briefly the history of club growth in the East Bay. The first Democratic club there was founded in 1934. As elsewhere, the impetus for more clubs came from Adlai Stevenson's presidential campaign in 1952. One estimate of their number is that, at the height of their activity, there were about eleven of them, 'each with about fifty to seventy-five members'.[32] These clubs, which were smaller than the ones in Los Angeles, were groups of like-minded people. Their disadvantage was that, while anyone could find a club whose members had interests similar to his own, a loose club structure was not well-suited to efficient electoral organization. This was recognized by some members as early as 1952, and certainly by mid-1953, following the liberals' failure in the Berkeley city council and school board elections, it was widely recognized. These defeats resulted in the creation of three organizations.

The first was the Berkeley Caucus (of Democratic Clubs) which was a body to co-ordinate club activity, and which made endorsements in Berkeley city elections. Caucus rules specified who qualified as a voting member of it and the procedures to be followed in endorsing candidates and in establishing an election platform. It was Berkeley based, and not Alameda County based, because essentially it was a vehicle for co-ordinating activity in local politics, and the services of city government were of more direct concern to the membership than those of county government. Its local orientation was emphasized in the Caucus constitution which stipulated that only Berkeley residents could be members. The second organization was the Eighteenth Assembly District Democratic Precinct Organization (or 18th ADDPO). This was first established for the 1954 congressional elections. It was a club in the sense that its members were supposed to pay dues, but these were only nominal and its purpose was 'un-club like'. The idea was that it would provide a forum in which one Democratic 'representative' from every precinct in the district could meet others to plan electoral activity. Unlike the clubs, it met only four times a year, though it did have one part-time employee. Many of its members were not members of other clubs, and they were supposed to be elected by Democratic voters in their precincts, not by clubs or their members. The 18th ADDPO was an attempt to create a party precinct organization like those found in the eastern cities, and to this end it was organized around the electoral boundaries of a partisan office, rather than being based on the city of Berkeley, as the Berkeley Caucus was.

The third organization to be created by the amateur Democrat movement was the Democratic party of the Seventh Congressional District (7th CD). The Eighteenth Assembly District fell within its boundaries, but the 7th CD also included the parts of Berkeley and Oakland which lay in Rumford's (Seventeenth) Assembly District. In effect, the amateurs gave life to an official party institution, a sub-structure of the state party, which otherwise would have been no more than a title on paper. The organization was significant in two ways. First, it had a permanent office which served as the headquarters for all party activity in the area. Mostly it was staffed by women volunteers and paid for by small subscriptions from individuals, although the clubs also made donations. It meant that there was a local focus for party work which prevented the growth of a wholly state-oriented view among local activists. Secondly, if the 18th ADDPO gave the whites some semblance of precinct organization, the 7th CD was the institution through which the coalition of blacks and whites worked. Having a separate organization minimized the tension that black inability to organize on the precinct level would have created. The 7th CD had two co-chairpersons, one of whom was always D. G. Gibson, while a succession of white women filled the other post. With a single individual in a formal institution to arrange co-operation on the black side, white politicians were able to stabilize white-black relations in the party.

It must be recognized, though, that even with these organizations campaigns for partisan offices were still not party-centred in the East Bay. In successfully contesting the Seventh Congressional District in 1958, Jeffery Cohelan did have access to the personnel of the 18th ADDPO and to some finance from it. Yet it was he who had decided to run for election – he had not been selected by the 18th ADDPO or by any other organization. He was responsible for running the campaign, and it was his close contacts with the labour unions which persuaded them to give money to the campaign. Moreover, although there was a liberal Democrat running for the Eighteenth Assembly District, Cohelan did not share any campaign literature with him or with any other candidate. The two campaigns did liaise with each other, and the candidates did attend some political meetings together, but the slate literature was the only device which had its origins in the party organizations, and there was much less of it than Cohelan's own literature. In other words, even at the height of its power, the liberal activist movement in the East Bay did not change the role of the candidate in the electoral process. What it did was to introduce party organizations into that process as an additional actor which a candidate could build into his electoral coalition; before then the Democratic party had had very limited involvement in the process. Real party organization had developed where previously there had been none; for a while, as elsewhere in California, it seemed as if party structures, as well as two-party competition, were reappearing in the state.

### 3. Denver: Party Politics as Usual

In contrast to the East Bay, the electoral role of the two parties changed very little in Denver between 1940 and 1960. The two main developments in this period were the collapse of a quasi-partisan machine centred on the office of mayor, and the growth in power of liberals in the Democratic party organization, at the expense of conservatives. If it was party politics as usual, in the Democratic party it was politics with an expanded activist base. The significance of the two developments is best explained by examining three features of Democratic party activity: nominating and campaigning activities for partisan legislative and executive offices, party involvement in city elections, and partisan judicial elections.

Unlike California, Colorado had highly competitive partisan elections in the 1940s. Denver was the centre of Democratic strength in the state, and in a 'Democratic year' the multi-member district system used for both legislative chambers would permit the Democrats to win every seat contested. This happened in 1948 when the fifteen General Assembly seats and four Senate seats were won by Democrats. In a 'Republican year' the Democratic monopoly would be broken by ticket-splitting – in 1950 the Democrats won only nine Assembly and two Senate seats. In normal circumstances the party's nominees were, in effect, chosen by the party officers and district captains, and 'normal circumstances' meant that there was no mass participation by political newcomers at the party caucuses. When dealing with known participants, most captains could have the people they wanted selected for the County Assembly. However, the power of the captains was really the power to determine which position a candidate would have on the primary ballot. A determined outsider might well obtain the votes of 20 per cent of the delegates, and thereby gain access to the ballot, but in a city-wide primary election for fifteen places a candidate was unlikely to succeed unless he had one of the first fifteen places on the ballot list. Because the order of the candidates on the list was fixed according to the number of votes obtained at the County Assembly, the district captains were in a position to defeat opponents. Occasionally, lower-designated candidates did defeat those who had a favourable position on the ballot, but mostly these losers were either extreme party mavericks, those involved in political scandals, or those widely perceived as incompetent.[33] Thus, County Assemblies were preceded by bargaining between would-be candidates and captains, and among the captains and party officers themselves. Although the candidates were often self-recruited, as opposed to being drafted by the party, the captains and officers often determined political career prospects.

In the primary elections the party organization was not a participant, but individual captains and committee members worked extensively for particular candidates. Despite the absence of the county party, a candidate who had not received a favourable position on the ballot list faced great difficulties in overcoming his initial disadvantage. He was running for a party-time office, with a constituency of over 400,000, against at least twenty-four or twenty-five opponents and sometimes many more. Nor could he devote all his campaign funds to making contact with the electorate, for the party required all primary candidates to contribute money to the county organization. Between 1940 and 1960 the amount demanded from each candidate increased from under $100 to about $175; the successful candidates in the primary then had to contribute about twice this amount after their place on the general election ballot was secured. The party had the power to do this because anyone who failed to contribute would not find favour with the party hierarchy in future nomination contests. All of the money raised, which was usually more than $5,000, was put towards the party's expenses at the general election; this formed one of the four main sources of the party's campaign funding – the others being the labour unions, a wealthy individual, and the city's congressman. Generally, the party did not use its nominating power to extract money from any other individuals, but there was one exception to this. In north-west Denver, then an area of traditional, 'ethnic', neighbourhoods, the district captain solicited funds from small businesses; this money went only to candidates whom the captain was supporting. It was a more common practice in eastern cities where the ethnic neighbourhoods were much larger, but its limited use in Denver is one of several indicators of the hybrid nature of party politics there.

Party control over nominations was also used to enforce another rule. The county organization required that, on all general election literature distributed by a candidate, the names of other Democratic candidates must be listed on the back.[34] The party-centred nature of campaigning is revealed also in the relatively low levels of expenditure by individual candidates. Most legislative candidates would spend only a few hundred dollars on their own campaigns, and while a few might spend more than $500, especially in state Senate campaigns, this was a small amount when compared with the $30,000 or $40,000 which the county party could be spending. In his general election campaign the candidate faced problems similar to those he faced at the primary in trying to do better than his colleagues. Even though his literature would now be distributed by all the active precinct committee members, it lost some of its impact because it was delivered with the literature of other candidates, and he had to meet party deadlines if it was to be delivered at all. Invariably he would have a few friends and political allies to help him, but normally the candidate did not have his own organizational structure; the loyalty of the precinct committee members was primarily to the party, and he would know very few of them personally so he could not persuade many of them to devote their energies to his campaign.

At the head of this rather powerful party organization was a leader who was more than a figurehead. In 1950 Lawrence Henry, a lawyer from the liberal wing of the party, became chairman and he served in the post for ten years, acting as a political broker in the same way as his conservative predecessors. He secured jobs for some party people, both in the public and private sectors; if he did not have access to the amount of patronage found in cities such as New York, it was sufficient to keep the party organization effective. It is difficult, if not impossible, to measure its effectiveness but the evidence is that, far from being in an advanced state of decline, the party's election activities were considerable; one estimate, provided in an interview, was that about three-quarters of all precinct committee members would have been involved in one activity, that of calling on those registered to vote during the last two hours of polling. The power of the chairman was bolstered by one source of party campaign funding. A wealthy businessman, from the conservative wing of the party, regularly guaranteed the chairman $15,000-$20,000 for the last week of the campaign. Although the money was raised by the businessman, the chairman was left to decide how it should be spent.

Under the direction of the party chairman, the party always had an extensive general election campaign. In addition to advertisements in newspapers and on billboards, it sent out a sample ballot to every registered voter, to indicate which were the Democratic candidates, and this was the main item of the party's expenditure. On election day many precinct committee members were paid $10 each to spend the day knocking on doors in a 'get out the vote' effort. Payment to them was necessary to prevent too many of them acting as paid judges at the voting booths, as this would have prevented them from canvassing voters. To supplement this party effort, some labour union members would take leave from work on election day to canvass voters; here again it was the party which could direct this help to precincts where it was most needed. This relatively centralized form of campaigning helped to influence the nature of the relationship between the city party and its congressman. He did not need the party's help to win re-election, for only in an exceptionally good year for the Republicans nationally might he face defeat. Yet poor relations with the party could generate sufficient opposition among district captains to bring about a serious challenge at the County Assembly.

In the 1950s this might have been a problem for Byron Rogers, who was first elected to Congress in 1950 after the incumbent liberal Democrat chose to run for a US Senate seat that year. Rogers was a well-known member of the party's conservative wing, and he won the seat in the same year that a liberal became party chairman. Liberal influence in the party increased in the 1950s, and Rogers might have been vulnerable had he not co-operated with the party. In fact, he avoided trouble by choosing not to distance himself from the party organization. During his twenty-year tenure in Congress he had only a small election organization of his own, spent only a few thousand dollars on each of his own campaigns, and associated his own re-election bids with the county party's campaign for the entire Democratic slate. To this end, he would raise several thousand dollars (usually between $3,000 and $5,000) from large businesses in the city which he would then donate to the county party. In the same way as the state legislature candidates, he received from the party an assessment of how much he should contribute, but at $2,000 this was less than the sum he actually donated. It can be argued that Rogers tied himself more closely to the party than would have been necessary for a candidate whose policy views were closer to those of the emerging liberal majority in the county party. On the other hand, relying on the party was inexpensive, and it meant that he did not have to worry about political organizaing at home in an era when paid regular flights to his district were not one of a congressman's perquisites.[35] That there was some flexibility in the relationship between party and congressman indicates again how the Denver party was a hybrid of western decentralization and East Coast centralization in local politics.

This point is also apparent when we consider the quasi-partisan machine in local government which lasted in the city until 1947. Under an early twentieth-century machine politician, a clause had been inserted into the city charter which required all municipal employees to serve at the pleasure of the mayor. The machine collapsed and there ensued a brief period of confusion and disorganization in municipal affairs, but the 'patronage clause' was not repealed. These years around 1920 were ones of highly volatile politics in the state, with radical labour union activity and the rising power of the Ku Klux Klan who succeeded in winning one gubernatorial election. Fearful of the future, a group of Republican businessmen supported a well-known, middle-aged Democrat in the mayoral election of 1923. This candidate, Ben Stapleton, was to remain in office for all but four of the next twenty-four years. (He was defeated in the 1931 election.) The business community continued to support Stapleton because of the stability his rule produced. Moreover, because he was seen to 'make the city work', as was Daley later in Chicago, Stapleton's methods did not provoke sustained opposition among Denver's middle-class voters. Thus can be explained an apparent paradox – the maintenance of a political machine in a state described by Daniel Elazar as having a 'moralistic' political culture.[36] But what exactly were the components of Stapleton's machine? One element was his access to campaign funds from the businenss community. Again, during election campaigns he could command the services of many city employees who 'served at his pleasure'. Here we must distinguish between the senior executives in the administration who were totally dependent on him, and who usually played major roles in the campaigns, and those in more menial jobs. The contribution of the latter was similar to the work they carried out in partisan elections, as precinct committee members, and this brings us to the complex status of the machine in relation to partisan politics.

The machine was the mayor's machine. Unlike the Daley case, this was not a case of a party leader using tenure in the mayor's office to recreate a *party* machine. Stapleton was not the leader of the county Democratic party, nor was he under its control; in addition to receiving campaign funds from Republicans, he also appointed some Republicans to positions in his administration. He could not be controlled by the Democratic party because his non-partisan office was not subject to the nominating procedures which made the party powerful in partisan elections. Equally, although he was a senior Democrat in the city, Stapleton did not run the party either directly or through surrogates. He had many employees and supporters in the party organization, but generally he remained aloof from its internal conflicts. City employees were too small a group to dominate the party organization, though they were a major element of its conservative wing. The personal nature of the machine, and its only partial identity with the Democratic party, is revealed in Stapleton's defeat in 1947. The election was won by a candidate who ran as a reformer and who, although backed by liberal Democrats, sought to characterize himself as having a broader coalition than that. His defeat of Stapleton did not bring about a rift in the county party nor represent a strong shift in the organization to the liberal wing. (When Lawrence Henry became party chairman three years later, Byron Rogers, an ally of Stapleton, received first-line designation for the congressional primary that year.) Perhaps the most telling point about the personalized character of the machine is that in 1947 Stapleton finished a poor third in the election; perceived as being too old at 77, as representing an out-dated style of politics, and as no longer providing the standards of services the city needed, he had no institutional resources, beyond the office itself, which he could command to contest the election seriously.

Once in office the new mayor, Quigg Newton, modified the city charter. Nevertheless, what remained was not an emasculated mayoralty, for the office-holder still had the power to appoint most of the senior executives in the administration without constraint. Moreover, the adoption of the civil service system did not make for more genuinely non-partisan competition for municipal offices. There were four main reasons for this. First, there would always be some jobs, including temporary ones, to which civil service rules could not be applied. Secondly,

Stapleton's removal from office did not make political out-casts of all his associates. Thirdly, the Stapleton years of rule had produced an ethos of political activism among municipal employees which could not be removed immediately by the changed conditions of employment. Finally, the remaining powers of the mayoral office still made it attractive to senior party politicians, and this facilitated party involvement in city elections. The mayoral elections of 1955 and 1959 were essentially contests between two candidates who had strong support from their respective party organizations, or, rather, in 1959 this was the character of the contest at the run-off election stage. Thus, the removal of Stapleton and the 'pleasure of the mayor' clause did not remove the partisan component from municipal elections; on the contrary, it can be argued that most of the elections between 1947 and 1967 were more overtly partisan than they had been before. This partisan nature of the contests led many Democrats to campaign for a change in the status of city elections. In 1960 they were able to make the issue the subject of a referendum, but it was defeated by a majority of 2:1.

A partisan electoral system would have been of considerable benefit to the emerging liberal majority in the upper echelons of the party organization, for they would have been able to neutralize two advantages conservative Democrats had in the existing system. One was their ability to appeal beyond the party to Republican-inclined voters and financial contributors, particularly when the Republican-backed candidate was fairly weak. With a partisan primary, Republican money would have been more likely to have been directed at candidates in that party's primary, and it would have been more difficult for conservative Democrats to have appealed to 'cross-over' voters. The second advantage was a legacy of the Stapleton era: the party had proportionately more conservatives at the precinct committee level (some of them city employees) than at the Executive Committee level. This gave the conservative candidate the campaign workers he needed to be competitive when the nomination was decided wholly by election; if it was effectively determined by the captains' ability to give a first-line designation in the primary, then the value of these workers would be reduced. In contrast to New York, where exclusion from power in the party could result in the growth of anti-party sentiments among liberals, we find in Denver that it is the newly-mobilized middle-class liberals who are among the strongest supporters of party structures in the 1950s.

In Denver, then, the politics of City Hall had only been partly linked to the politics of the party organization before the 1950s. This was not so with the judiciary. At the county level both the district attorney and the district court judges were elected on a partisan ballot. This had two main consequences. It drew lawyers, especially ambitious ones, directly into party politics. Of course, electoral politics has always been a means for American lawyers to advance their careers, but a partisan ballot made party organizations the central arena for their participation. For the parties the connection provided both a pool of political talent and a source of campaign funds. Furthermore, a partisan ballot for judicial posts also provided a source of patronage. Judges employ bailiffs and clerks and in Denver, as in many cities, these offices were often filled by district captains as a reward for nomination support and campaign activity. In this process the county party chairman had a crucial role, for it was usually through his office that discussions about who might need a vacant job were conducted. Estimates of how many district captains were court employees vary, but it is possible that in the 1950s as many as one half of them were.[37] This patronage consolidated the position of the chairman and the role of the party as an electoral intermediary.[38] Because it did not affect their policy objectives, nor their ability to achieve them through the party, the Denver liberals were content to leave judicial patronage as it was; it was not an issue which divided them from the conservative wing of the party. Once again, this provides a contrast with New York City where dismantling judicial patronage in Manhattan was to be one of the few direct successes for the reformers. There it had been a device which helped to keep in office those who were denying influence to the liberals in the party.

These brief accounts of change in the Democratic parties in the three areas between the New Deal and the early 1960s reveal very different developments. In New York City we find internal division in the party which was exacerbated by the rise of the amateur club movement. Under De Sapio in the 1950s, the party regulars in Manhattan failed to provide unified leadership and to revitalize the party, in the way that Daley did in the Chicago party. The electoral débâcle which followed the Buffalo convention in 1958 put the regulars in an even worse position, for they were now less able to withstand challenges from candidates who sought to mobilize long-standing 'reform' sentiments against 'machines'. In New York City, then, there is evidence of obvious and substantial party decay preceding the collapse of the 1960s and 1970s. This was not evident in either the East Bay or Denver where the party organizations thrived in the 1950s. In the East Bay the party-oriented amateur activists created party structures where previously there had been none, and in Denver liberals co-exited peacefully with the party's conservatives while they were replacing them as the dominant group in the party. But to which of these 'models' of change in the Democratic party would we be advised to turn when attempting to assess what was happening nationally? Those who accept James Q. Wilson's version see New York as typical of the potential conflict which was building up in the parties; some observers, indeed, have seen the intra-party confrontations of the later 1960s as a continuation of the earlier skirmishes. However, it can be argued that neither New York nor Chicago were typical, because in these cities so much patronage remained to be fought over in the 1950s that regular politicians would not acquiesce in the sharing of power with reformers. In turn this made patronage a central target for the amateurs. In many cities, though, the patronage resources of the machines, and the machines themselves, were already declining by the end of the Second World War. Accommodations with the remaining vestiges of patronage-based organizations were easier for the liberals; in these cities a more orderly, and gradual, transformation of the party organizations could emerge. If most parties were not experiencing the kind of organizational revival of the East Bay, then neither do most of them seem to have experienced bitter internal conflicts, or crises in their ability to perform electoral tasks. In many places the Democratic organizations were, probably, a mixture of quietly decomposing patronage operations and small upsurges of middle-class enthusiasm for politics; given what was required they could perform campaign tasks adequately. Now this is not to deny that the intense issue conflicts of the mid-1960s would strengthen anti-party sentiments among some of the new participants. But it is misleading to suggest that the amateur politics of the later period was merely the amateurism of the 1950s on a larger scale. If amateur politics is seen as a 'time bomb' waiting to explode under the Democratic party, then it must be added that it was a 'bomb' missing a number of components. In the peculiar circumstances of politics in the 1960s these components were produced, and the Democratic party faced a crisis from within. Nevertheless, this was only one of the factors which was to end the short 'Indian Summer' which the parties had enjoyed in the 1950s; there were other forces which were already undermining them.

## Notes

[1] See, for example, Sayre and Kaufman, *Governing New York City*; Costikyan, *Behind Closed Doors*; Wilson, *The Amateur Democrat*; Theodore J. Lowi, *At the*

*Pleasure of the Mayor* (New York: Free Press, 1964). The amount of research which has been conducted on various aspects of New York City politics is enormous; some of the published, and even unpublished, papers are cited in the footnotes to chapters 5, 6, and 7 of Ira Katznelson, *City Trenches* (New York: Pantheon, 1981). There has been much less research on East Bay politics, though there was a major project on public policy-making in Oakland (The Oakland Project) carried out under the auspices of the University of California in the late 1960s and 1970s. There has been only a very limited amount of research on politics in Denver.

[2] Morris P. Fiorina, 'The Decline of Collective Responsibility in American Politics', *Daedalus*, 109 (1980), 29.

[3] V. O. Key, *American State Politics* (New York: Knopf, 1956), p. 71.

[4] Jewell and Olson, *American State Political Parties and Elections*, p. 32. States with non-partisan legislatures have been excluded from our discussion.

[5] Jewell and Olson, *American State Political Parties and Elections*, p. 33.

[6] One of these four instances, Maine, involved an Independent winning the governorship.

[7] Austin Ranney, 'Parties in State Politics', in Herbert Jacob and Kenneth N. Vines (eds.), *Politics in the American States*, 3rd edn. (Boston: Little, Brown, 1976).

[8] Robert H. Salisbury, 'The Urban Party Organization Member', *Public Opinion Quarterly*, 29 (1965–6), 550–64; Samuel J. Eldersveld, *Political Parties: A Behavioural Analysis* (Chicago: Rand McNally, 1968), 149–50.

[9] Wilson, *The Amateur Democrat*, p. 37, citing Sayre and Kaufman, *Governing New York City*, pp. 688–9.

[10] Katznelson, *City Trenches*, p. 122.

[11] Shefter, 'Regional Receptivity to Reform: The Legacy of the Progressive Era', p. 480.

[12] On the reduced role of party in recruitment to the élite positions in New York City administrations, see Lowi, *At the Pleasure of the Mayor*, especially Chapters 4 and 5.

[13] Gifford, 'New York City and Cosmopolitan Liberalism', *Political Science Quarterly*, 93 (1978–9), 576.

[14] Susan Tolchin and Martin Tolchin, 'How Judgeships Get Bought', *New York Magazine*, 15 Mar. 1971.

[15] The phrase is taken from Sayre and Kaufman, *Governing New York City*, p. 186.

[16] Wilson, *The Amateur Democrat*, pp. 44–8.

[17] There are numerous accounts of this; see, for example, Costikyan, *Behind Closed Doors*, Chapter 12.

[18] Wilson, *The Amateur Democrat*, pp. 32–6.

[19] These figures are taken from Jewell and Olson, *American State Political Parties and Elections*, pp. 333–7.

[20] Dan Nimmo, *The Political Persuaders* (Englewood Cliffs, NJ: Prentice-Hall, 1970), pp. 35–6.

[21] Interview no. 49.

[22] Interview no. 12.

[23] Interview nos. 12 and 46.

[24] Jeffrey L. Pressman, 'Preconditions of Mayoral Leadership', *American Political Science Review*, 66 (1972), 516.

[25] An account of these changes is presented in a longer, and unpublished, version of an essay by T. J. Kent, Jr. The published version is 'Berkeley's First Liberal Democratic Regime, 1961–1970'.

[26] See the set of short essays collected by Evelio Grillo, 'D. G. Gibson: A Black who Led the People and Built the Democratic Party in the East Bay', in Nathan and Scott, *Experiment and Change in Berkeley*.

[27] Interview nos. 16 and 32.

[28] Wilson, *The Amateur Democrat*, p. 125.

[29] Eugene C. Lee, *The Politics of Nonpartisanship* (Berkeley: University of California Press, 1960), p. 36.

[30] Wilson, *The Amateur Democrat*, p. 125.

[31] Ibid., p. 122.

[32] Kent, 'Berkeley's First Liberal Democratic Regime', p. 80.

[33] Interview no. 104.

[34] Interview no. 109.

[35] That Rogers's strategy was not the only one open to a Democratic congressman in the 1950s can be shown by comparing Rogers with his predecessor, John Carroll. The liberal Carroll was first elected in the Republican landslide of 1946, when he defeated a Republican incumbent. In doing this, it was estimated that he spent $20,000 – a large sum by the standards of the 1940s – while before 1970 Rogers never spent more than $10,000. Carroll had shown that a high-spending, relatively independent campaign was possible in Denver, even before modern campaigning techniques were introduced. Undoubtedly, with views more compatible with the party's liberal majority, he could more easily have campaigned independently of the party organization in the 1950s than Rogers.

[36] Daniel J. Elazar, *American Federalism*, 2nd edn. (New York: Crowell, 1972), ch. 4.

[37] Interview no. 84.

[38] Interview no. 109.

# 4
# The Missing Party Work-force

If it is to perform election campaign tasks, a caucus-cadre party, like any party must recruit participants to perform them. The number of activists a party requires depends on several factors, including the number of voters within its territory, the techniques available for stimulating voter support, the relative cost of those techniques, and the political geography of the territory. Although opinion polling, the use of computers in direct mailing to voters, and television advertising reduced the size of the workforce required for certain kinds of campaign activity in America, most campaigns in the 1970s still required more activists than could be raised by the individual candidate's personal contacts. In the mid-nineteenth century this consideration had been the key to the growth of party organizations in a party structure based on small groups of notables. The parties had access to sources of labour for the election tasks, and this was both an attraction for would-be office-seekers and also the basis of the notables' ability to control elections, through their control of the nomination process. From the middle of the twentieth century, the Democratic party organizations' role as oligopoly suppliers of electoral services was threatened in a number of ways. One of the problems they faced, though, was somewhat older in origin and was not an *immediate* threat to party-centred politics – this was the difficulties they encountered in recruiting activists. In examining this in this chapter we utilize the much discussed distinction between 'professional' and 'amateur' activists.

Underlying this distinction there is a further distinction between the kinds of reasons people have for participating in politics. James Q. Wilson conceived participants as responding to three possible incentives – which he called material, solidary, and purposive incentives.[1] Later, he was to introduce a further distinction between *specific* solidary and *collective* solidary goals.[2] The central idea of the earlier classification is rather simple but, regrettably, it came to be discussed in confusing and ambiguous ways. A material goal has as its objective something which is immediately available to a person, is enjoyable by him alone, rather than as a member of some collectivity, and is a good of a kind usually exchanged in economic markets. Obvious examples are money, jobs, or a food hamper at Christmas. The object of a solidary goal is immediately available, could be of a kind enjoyable either by the person himself or only when he is part of a collectivity, and is not exchangeable in an economic market. Examples of these objectives include prestige and the enjoyment of participating with other people. A purposive objective is one which is realizable only in the long term, and not immediately, is one which will affect people other than those pursuing it, and may, but does not necessarily, relate to goods exchanged in economic markets. Examples would include activity to promote a full employment economy, the realization of the Kingdom of God on earth, or the ending of the death penalty.

The obvious point to make about this classification is that it combines several different dimensions – who is to benefit (the individual, a collective, or both?), is the benefit an economic good, and is the goal realizable immediately or only in the longer term? The 'looseness' of this classificatory system does not matter because it does seem to separate in a useful way the main goals participants have been pursuing in contemporary western polities. The 'looseness' only becomes a problem if we attempt to 'squeeze too much out' of the distinction and, in particular, if we assume that all possible objectives can be fitted neatly into the three categories. Unfortunately, Wilson himself made this mistake. For example, he sees the purposive incentives an organization might offer as all being similar to opposition to capital punishment – they involve a 'worthwhile cause' in which, at most, the individual is only an indirect beneficiary of the good's realization.[3] This would mean that, to the extent that a working-class man became a socialist in order to pursue his own interests, we should have to say he was not pursuing a purposive goal. The fact is that we can only narrow the category of purposive incentive in the way attempted by Wilson, by devising a schema which is far more complex than this one and takes account of the several dimensions involved. This is unnecessary for our purposes, because none of the data collected by political scientists on the goals of party activists is detailed enough to make it worth while. The three-unit classification devised by Wilson is adequate providing we remember its limitations.

Having set up this framework, Wilson initially seemed to be making a distinction between amateurs and professionals on the following lines. Amateur activists were attracted to politics solely by purposive incentives, while the effective incentives for professionals were either material or solidary ones.[4] The major disadvantage of this was that, while it isolated as 'amateurs' a particular kind of activist, it made the category of 'professional' far too broad to permit a satisfactory analysis of trends in activism in America'a political parties. The picture that emerged in Wilson's study was a distorted one, of an influx of a new kind of participant in the 1950s (his 'amateurs') who lay outside the mainstream of American party politics. In fact, those who were mobilized partly by purposive incentives in the 1950s seem to have been a rather diverse group.[5] They included people who sought solidary, and sometimes material, goals as well as public policy objectives. The amateur movements Wilson identified were only partly 'amateur' in his sense of the term. It is, therefore, more useful to conceive the amateur–professional distinction as involving a spectrum; at the centre of it are those who had only solidary objectives and those for whom participation was the result of a mix of material, solidary, and purposive goals. In talking about amateurs and professionals, we are talking about two tendencies on each side of the spectrum, and we are recognizing that only some activists will represent pure instances of each type. It is in this way that we use the two terms here.

There are two further problems with the Wilson analysis which are worth mentioning. As Hofstetter has argued, Wilson constructed two ideal-types of politician who supposedly differed in a number of their attitudes towards politics.[6] However, the willingness to take electoral risks, supposedly one of the main characteristics of amateurism, has been shown by subsequent researchers to be linked to whether a group is an 'in' or 'out' group in the power structure of a party, rather than a public policy orientation.[7] A second problem is that the complexity of the Wilson ideal-types has produced even more confusion among political scientists as to the basis of the distinction between amateur and professional. Some of them have seen it as reflecting a difference in attitude towards risk-taking in the electoral process. Professionals believe 'elections should be fought to win' and that decisions should be made by '"weighing" interests of community and party'.[8] They are risk-averters, unwilling to risk electoral loss even if, in doing so, there is an opportunity for securing substantially more benefit should they succeed. On this account the amateur is Wildavsky's 'purist' and the professional his 'politician'.[9] Further difficulties have arisen because the term 'amateur' is sometimes used to refer to advocates of procedural democracy in political parties, though on this view it would exclude many of the Goldwater activists; and there are other uses of the term besides this.[10]

Having explained how the terms 'amateur' and 'professional' are employed in this book, we can turn to examine why the work-force of the Democratic parties has been reduced. In the first section of this chapter we discuss some of the general social changes which have limited the supply of material, purposive, and solidary incentives to the parties. Then, in the second and third sections, we consider how the recruitment of professional and amateur activists became difficult in the three areas of our study.

## 1. The Problems of Mobilizing a Party Work-force

### *A. Material Incentives*

Until the late 1960s it was universally agreed that, at least since the New Deal, the recruitment of party activists through material incentives had declined, and that those who remained in the parties were less controllable than their predecessors had been fifty years ealier. It was further recognized that falling recruitment was one of the main causes of the reduced power of America's urban political machines. Various reasons were given for this failure of the machines to recruit a satisfactory work-force. The supply of legal immigrants had been restricted after the early 1920s, thereby eliminating the main source of poor and socially unintegrated people who had been willing to play 'the rules of the game', in return for a patronage job from the party. Again, the prosperity which followed the Second World War made these low-paid, and mainly unskilled, jobs less attractive, and this deprived the party organizations of much choice in deciding to whom to allocate a job. The extension of the civil service system further reduced both the number of patronage posts in local government and also the opportunity for a party to use the threat of dismissal against those whose party work was inadequate. Moreover, increased wage levels meant that the kind of payments traditionally made for the distribution of campaign literature, or for canvassing on election day, were simply too small to be an effective inducement. Hence, it was widely assumed that the most distinctive feature of American party politics, the hierarchical political machine organized around 'materialist' activists, was gradually being eliminated from the cities.

In the early 1970s a small, but influential, so-called 'revisionist' argument developed which modified, though it did not replace, the conventional wisdom.[11] There were six main elements in the revisionist viewpoint. First, studies of ethnic groups, the principal electoral base of the machines, had demonstrated that the Irish, Poles, Italians, and the rest were not as fully disaggregated in American society as was earlier believed. Ethnic loyalties and ties to symbols of ethnic identity, such as churches, remained even though migration from old neighbourhoods to new suburbs had destroyed the physical proximity which members of these groups had enjoyed.[12] Secondly, it was argued that, if there was a decline in the number of patronage jobs in the eastern and mid-western cities, in some of them it was a minor decline. Not only had Chicago avoided the national trend but, according to Raymond Wolfinger, even in cities such as New Haven and New York there were still many patronage jobs available. Thirdly, the historian Bruce Stave demonstrated that, at least in Pittsburgh, the jobs created under the New Deal programmes had helped to bolster the city machine. Fourthly, the expansion of government services at all levels since the 1930s had also increased the opportunity for public officials to reward supportive businesses with government contracts; the politics of preferment had arguably increased, rather than declined. Fifthly, government expansion had multiplied the number of policy-making and skilled jobs at the disposal of the President, state Governors, city mayors, and others, and that these were not usually covered by civil service rules. Finally, more government contact with the individual citizen had actually increased the need for organizations to act as 'trouble-shooters' on behalf of aggrieved or helpless persons.

Many of the points made by the revisionists are correct: America is not an ethnic 'melting-pot', opportunities for venality and preferment in public life remain high and might have increased, and people do seek out help in dealing with government bureaucracies. But these matters are only peripheral to the central question about political machines and their maintenance as electoral intermediaries, which is, Has there been a substantial reduction in their ability to use jobs and favours, as a means of extracting work for the party, from those who are the beneficiaries? It is not simply the number of jobs and favours involved, which is at issue, but the ability of the party to maintain the amount of political activity it 'receives' from those it assists. As we shall see, the evidence suggests that few parties may have been able to maintain the size or quality of their 'materialist' work-force. While this does not diminish the value of the revisionist argument, it does throw doubt on those who would use it to claim that political machines were not in decline as electoral intermediaries in the 1960s.

### *B. Purposive Incentives*

The level of purposive participation in a political party depends primarily on three factors: the time and money needed to afford such activities, the events or people that stimulate concern for particular interests, values, and ideals, and the alternative institutions or processes through which the objectives can be sought. In relation to these factors, we propose two relatively uncontroversial hypotheses.

The first is that the promotion of 'long-term' collective goods in the western world is *normally* practiced only by those who have sufficient free time and by those who have a permanent source of income which they regard as adequate. We would stress the qualification, 'normally', in this claim, for religious beliefs, ideologies, peer-group solidarity, charismatic leaders and so on do stimulate participation by the poor and those with relatively little spare time. Nevertheless, we would expect that increases in both spare time and income would be associated with the rise of movements based on purposive participation. Not surprisingly, then, we find in the four decades after the end of the Second World War the growth of many political movements centred on the middle classes: the amateur movement in the Democratic party, and later in the Republican party, and movements based on pressure groups embracing policies such as the Vietnam war, environmentalism, abortion, women's rights, local and state taxes and many others. After 1945, there were for the first time millions of people with sufficiently large incomes and hours of leisure to devote to the pursuit of goals other than their immediate self-interest. Indeed, within a few years most Americans would be in a position to 'afford' purposive participation, though only a relatively small minority would choose to engage in politics.[13] Differences in participation rates are related, of course, to differences in education, social class and so on – issues central to political science but not to our study. The point we must emphasize is that the Second World War was a turning-point for purposive participation in America; there was a potential for change in the balance of participation found in the parties which did not exist before.

The second hypothesis is that the stimuli to this form of participation will vary over time. Of course, in admitting this we do not have to embrace theories such as that of Hirschman which seek to explain a supposed cycle involving the pursuit of private ends and public affairs.[14] But that there are periods of intense interest in public affairs followed by periods of relative quiescence among the citizens of liberal democracies seems undeniable. Since 1945 there have been two periods when an upsurge of policy-oriented activism has affected the Democratic party. The first lasted from 1952 to about 1954. The catalyst for this was the appeal Adlai Stevenson had among middle-class liberals in the 1952 presidential election campaign, and it was more Stevenson's political style than any particular issue which prompted this mobilization. However, the movement retained strength for several years afterwards, partly because it did not rely exclusively on purposive participation. The second period of mobilization lasted from about 1966 to 1974 and centred initially on opposition to the Vietnam war. Intense conflict over this issue within the American public, combined with the continuing conflict over civil rights, provided a catalyst for mobilization around other issues, including environmental, 'good government', and consumer problems. Once again, particular election campaigns, most notably those of McCarthy in 1968 and McGovern in 1972, promoted public concern with the issues and helped to focus attention on the Democratic party as one of the channels through which public policy objectives might be attained. However, linked to a tradition of 'anti-partyism' in America, policy-oriented participation could be a destructive force for a caucus-cadre party. Indeed, some accounts of the post-1968 reforms of the Democratic party's presidential nominating procedures stress the critical role played by this element of the activist movement in effectively dismantling the presidential party. Nevertheless, it must be emphasized that purposive activism need not be anti-party in character. As we have argued, many of the issue-oriented Democrats mobilized in the 1950s were very supportive of parties as electoral intermediaries, and a study of twenty-two state conventions in 1980 by Abramowitz *et al.* found that, although most activists were issue-oriented, party attachments were strong.[15]

### C. *Solidary Incentives*

The ability of the American parties to attract 'solidaristic' activists has been threatened by two developments in the last twenty years. The first is the decline in public esteem for the parties. Individuals who seek status or respect are less likely to be attracted by an institution which is held in low esteem, and opinion polls have revealed consistently that parties have become among the least revered of American institutions.[16] The second development has affected recruitment among those who might find enjoyment in the activities of party organizations. New technology, improved living standards, and changed social values have led to a vast increase in the alternatives people face when deciding how to spend their leisure time. At first, the general increase in leisure time after 1945 had a positive impact on the Democratic party – it was one of the factors which made possible the Stevenson-inspired mobilization of activists. However, the early 1950s can be seen as an important watershed in the social history of the American middle classes. Many of the technological and economic changes which reduced the amount of time they had to spend in employment or in housework had already occurred. The middle classes had more time for social activity than before the war, but they lacked the socially acceptable ways of meeting others which were to exist from the 1960s onwards. Political discussion and organization was conspicuous as a channel for their energies.

Of special significance in the Democratic party was the role of women, in terms both of their numbers and their contributions to the labour-intensive activities of the clubs and the formal party organizations. In some places the availability of a female work-force, and the absence of men to perform the tasks, meant that women even became the recipients of patronage which traditionally had been awarded to men. This was especially true of part-time campaign jobs. For example, in 1962 in the middle-class city of Denver, it was estimated that 90 per cent of the 2,000 election judges were women. They were the only people with sufficient time to spend at the polls; the jobs were allocated to faithful election workers by the precinct committee members, although their wages were paid by the Election Commission, and not the parties.[17]

Yet the middle classes were not the permanent source of activists for the parties that they might have appeared to have been at the time, for there were two social changes in the post-war world which were undermining their potential contribution. One was the revolution in beliefs about appropriate ways of meeting and socializing with others. Obviously, such innovations as the 'singles bar' are not the cause of the decline in party organizations, but they are manifestations of the changes in social values, and hence of opportunities, which have curtailed the growth of a middle-class activist base for the party. Equally significant has been the large increase in female employment and in single-parent families. This has drastically reduced a major source of recruits for the party, because both the employed and single parents have less time for social activities outside the home. From Tables 4.1 and 4.2 we can see not only the magnitude of the growth in female employment, but also one potential, if temporary, source for a middle-class activist base in the

**Table 4.1**: Female Labour Force as a percentage of the total US Female Population

| Year | % |
|---|---|
| 1940 | 27.4 |
| 1950 | 31.4 |
| 1960 | 34.8 |
| 1970 | 42.6 |
| 1979 | 50.7 |

*Source: Statistical Abstract of the United States, 1980,* Washington DC: Department of Commerce, Bureau of Census, 1980, p. 402.

**Table 4.2**: Married Women in the Labour Force as a percentage of total US Married Female Population

| Year | % |
|---|---|
| 1940 | 16.7 |
| 1950 | 24.8 |
| 1960 | 31.7 |
| 1970 | 41.4 |
| 1979 | 50.0 |

*Source: Statistical Abstract of the United States, 1980,* Washington, DC: Department of Commerce, Bureau of Census, 1980, p. 402.

Democratic party in the 1950s – married women. Although the proportion of women in the labour force increased in the 1940s and 1950s, the rate of increase at that time was only about half that evident in the following decades. Moreover, although there is much less difference between these two twenty-year periods with respect to the growth in the proportion of married women in the labour force, three married women in four were still outside the labour force at the time of Stevenson's first presidential campaign; by 1979 only one half of them were. For a very brief period circumstances were favourable to the growth of solidaristic participation among social groups which generally had not been part of machine politics.

Just as Wilson in *The Amateur Democrat* pointed to change in the social composition of the cities as an important factor in explaining the rise of a new kind of activist, so too we must point to social changes at the time Wilson was writing when seeking to explain why the amateur movement did not persist. The decline of solidaristic incentives in the 1960s and 1970s hurt the amateur Democrats as much as it hurt the party professionals, because it occurred so soon after the amateur movements had become established. This point is crucial in developing a full understanding of the problem of recruitment to the parties. Nevertheless, because there are virtually no data relevant to the matter, it is not something which we can explore in the next two sections. Any account of the decline of both professionals and amateurs would be incomplete, however, if it did not draw attention to the fact that by the 1970s parties had lost out to other social institutions as arenas for social gatherings.

## 2. The Disappearance of the Professional Activist

As John Harrigan has pointed out: 'The years immediately before and after World War II stand as a symbolic period in urban political history. One after another, several famed city machines lost their holds on their electorates.'[18] He cites the fall of the machines in Kansas City, Jersey City, and Memphis as instances of this, and we have already noted the collapse of the quasi-partisan machine in Denver. As in many of the other cities which experienced this transformation, the source of boss control in Denver, patronage jobs, was much reduced through the introduction of civil service rules in local government. In Denver the only sources of patronage after Quigg Newton's reforms were high-level administrative posts, temporary jobs, and jobs in the court system. The court jobs disappeared after 1966 when partisan elections for judgeships were abolished. Until then this patronage had remained a major incentive for recruitment to élite levels of the party organization. While in the early 1960s nearly half of the city's district captains might have been employed as court clerks, ten years later only one or two of these employees remained in party office.[19] The absence of many material incentives for activism in the Denver party in the later 1970s was also revealed in responses to one of the survey questions. Both the Executive Committee members and the precinct committee members were asked how important the following reason was for their continued participation in the Denver Democratic party: 'It provides me with an opportunity to meet the sort of people who might help me in my job or career'. Only 32 per cent of the Executive Committee (n = 51) and 19 per cent of the precinct committee members (n = 198) agreed that it was of any importance at all; only 6 per cent and 7 per cent respectively of the two groups claimed that it was a 'very important' reason for them. Thus, even considering material incentives in the widest possible terms, the Denver party does not seem to have been well endowed by the end of the 1970s. However, as revisionists might point out, in other cities the facilities with which a professional party workforce could be recruited were probably not reduced quite as dramatically as in Denver.

The revisionist arguments can be both useful and misleading in discussions of urban Democratic parties. They are useful because they do draw attention to differences between cities in the decline of patronage jobs at the disposal of the party organizations. Clearly, a few centralized patronage operations have survived. The major problem in evaluating the revisionist hypothesis, that the reduction in patronage has been greatly exaggerated, is that we cannot tell how prevalent the recruitment of professional activists has been without intensive studies conducted in a number of cities. There have been very few such studies, and this makes it difficult to put the exceptional cases of strong machine survival, such as in Chicago, into context. Too often the Chicago experience from 1955 to 1976 is seen as resembling that elsewhere, rather than as being highly unusual. One of the few studies of machines in recent years is Thomas Guterbock's research in Chicago, but he too fails to explain adequately how this case relates to others. He claims that 'the Chicago machine is not nearly so anachronistic as casual observation would suggest. In fact, patronage-based party organizations continue to play an important role in many large cities and many more smaller ones.'[20] But, in the supporting footnote when he quotes from an article by Greenstein written sixteen years earlier, he acknowledges that we simply do not know the distribution of types of party. While it is certainly the case that patronage parties still function in part in many cities, it is not widely accepted that most of these cities come close to approaching Chicago in respect of the comprehensiveness of *party* control. Guterbock has concealed the issue by his use of the vague expression 'important role'.

Revisionism can be misleading. It is if we fail to distinguish the survival of party-controlled patronage jobs from other resources which permit the practice of materialistic politics. One obvious point about both the politics of preferment and the expansion of non-civil service policy-making jobs is that, while they represent an increase in material incentives, they do not necessarily benefit the parties. Indeed, it seems more plausible to argue that these resources are precisely the sort which will be of benefit to the office-seekers individually, and will help them to loosen their ties to the party. This point will become clear as we examine what has happened to party control of patronage in New York City. Along with New Haven, Philadelphia, and the cities of upstate New York, New York City is often cited as an example of surviving patronage politics of a kind closest to the Chicago model.[21]

Far from declining in the 1960s, patronage at City Hall under Mayor John Lindsay actually seems to have increased. Wolfinger calculated that the number of 'provisional' employees increased eight times and, in terms of both the number of patronage jobs and the size of the patronage payroll, it seems that Lindsay exceeded his predecessor Robert Wagner.[22] The main vehicles for the distribution of patronage were the John V. Lindsay Associations, political clubs which arose once he was in office. But there is another way of looking at the Lindsay mayorality. While he, like his successor Abraham Beame, could provide $20,000-a-year bulldozer driving jobs to those whom his advisers chose, these were not typical of the resources at his command. Of 245,000 city jobs, only 9,000 were not covered by civil service regulations. Few of the non-civil service jobs were of a kind which could be used to recruit loyal, long-serving, low-level political activists. They were either genuinely temporary jobs, rather than ones which were permanently renewable, or they were high-paid, skilled jobs which served as rewards for his élite supporters. The main source of Lindsay's power was not so much jobs, but financial inducements; he practised what is often called the 'politics of preferment', rather than patronage politics *per se*.[23] Lindsay would reward financial backers through such devices as the right to select the banks where the city would place its funds. The principal Lindsay innovation was the extensive use of consultants; in his administrations expenditures on consultancy rose from $8m. to $75m. Some, though obviously not all, of these expenditures could be used as rewards for 'political services' and as inducements to provide the services in the future. Other kinds of contracting could be used in the same way. Perhaps the most bizarre revelation of the Lindsay years was that the city purchased $100,000 worth of ketchup a year from the H. J. Heinz company – Heinz having been a contributor to the Lindsay campaigns. Mayor Beame reduced the amount spent annually to $43,000, and he divided the contract among several companies. (What happened to the ketchup 'lake' which Lindsay created has not been revealed.)

Of course, it might be argued, Lindsay had to emphasize preferment politics because he had no party base. He had never been close to the

Republican organizations in the city, and he even failed to obtain the party's nomination in his re-election bid in 1969; the Liberal party, on whose ballot line he ran in both 1965 and 1969, lacked an extensive club structure.[24] Nevertheless, when Abraham Beame, an archetypal Democratic clubhouse politician, took over the office in 1973, it was impossible for him to build up a large, old-style professional party organization. Beame was able to exert some old-fashioned patronage power – he drove Queens party leader Matthew Troy from office – but there were neither enough non-civil service jobs, nor sufficient flexibility in the civil service rules, to increase party recruitment at the mass level. Or, to be more precise on the first point, there were not enough patronage jobs controlled by the mayor or the county party leaders which were at salary levels to attract applicants. In 1971 Brooklyn party leader Meade Esposito had complained about the quality of the jobs at his disposal: 'What are you going to do with a $7,000 job? It barely pays the rent.' Nor was he alone in taking this view. Again in 1971, a leader of a reform club which had only two members with patronage jobs explained: 'We've got plenty of people who would be glad to be commissioners, but they are not interested in an $8,000 clerk's job.'[25]

Apart from court patronage, which we consider as a separate category, what was available in the 1970s to keep materialistic rank-and-file activists content were income supplements, rather than full-time jobs. The highly paid bulldozer drivers mentioned earlier were well-publicized cases, as were the posts of pot hole inspectors, and the better paid jobs in the city's Board of Elections. There were too few of these jobs to maintain the county organizations in five boroughs. For the higher echelon club activists, more typical examples of remuneration for party work were the $3,750 'no show' jobs provided by money which Brooklyn Democrats in the state legislature were supposed to give to their district leaders from their staff allocations. For the lower echelon activists, there were jobs on election day which paid between $15 and $25.[26] This kind of part-time money could hardly be expected to induce the level of enthusiasm by recipients necessary to maintain continual and effective party activism. It was scarcely surprising, therefore, that just before the 1973 city elections Meade Esposito estimated that only about 25 per cent of the Democratic district captains were working at their jobs of distributing literature and 'getting out' Democratic voters. Moreover, the combined effects of insurgence and the decline in patronage jobs even for the party élite meant that, by 1979, a substantial minority of Brooklyn's forty-four district leaders did not hold city, state or county offices.[27] This was in a county organization that some journalists still referred to as one of the most powerful outside Cook County, Illinois.

To summarize the argument about patronage in New York thus far: the mayor of the city remained a position from which patronage jobs and contracts could be dispensed. For this reason the county parties did not cease their involvement in the elections for city offices. However, many jobs which could be filled at 'the pleasure of the mayor' were policy-making jobs requiring expertise more than personal or party loyalty. Neither the mayor nor the party leaders had a large supply of lower level jobs for which they could get applicants. While there were loopholes in the system of civil service rules, especially with continuing temporary jobs to which the rules did not apply, this did not generate sufficient jobs to recruit an army of activists.

How different, then, is New York from other cities in its failing patronage base? Any answer to this is necessarily speculative because we have only incomplete information both about New York and other cities. Patronage has a bad name and, while politicians sometimes speak freely about it when they lack it, it rarely happens that the newspaper reporter or the political scientist is in a position to extract much information from those who can dispense it.[28] From what we do know, it appears that there are more cities like New York than like Chicago, where in the mid-1970s each ward committeeman controlled between 100 and 400 jobs.[29] Interestingly, Guterbock's study of one ward in that city revealed that 'Ninety-seven per cent of the club's 150-odd members hold patronage jobs' for which the median salary in the mid-1970s was $8,000.[30] This is indeed a puzzle. How was America's second largest city able to support a thriving, hierarchical machine, while in its largest city the patronage system had become decentralized and no recruits could be found for jobs with salaries at the level of those in the Chicago party? Obviously, this raises the question, Why did Chicago seem exceptional? Unfortunately, the Guterbock study is unclear about this.

One relevant dissimilarity between Chicago and New York is the cost of living. This was higher in New York, as were the levels of welfare payments, and both of these considerations made $7,000 or $8,000 a year jobs relatively less attractive than in Chicago. But these differences are not so great that they provide an adequate explanation of the great variation in patronage recruitment to the parties. More important, perhaps, is the recognition that self-interest on the part of activists could not, of itself, account for the participation of the Chicago club members. This is a point emphasized by Guterbock: 'With avenues for advancement limited, and with the larger community holding club members in no special esteem, most of the club members have limited stakes in the party's future, and therefore will not reliably contribute to its efforts out of self-interest alone.'[31] It is at this stage that Guterbock's argument lacks clarity. The reader is not told directly whether it was additional patronage benefits, a mistaken belief by the club members that they could advance in the party structure through building voter support in the precincts, or the acceptance of values other than self-interest which kept the members active.[32] However, the main thrust of his argument seems to be that once they had decided to accept a low-paying job in the party, and had been socialized into the club, the club members identified with the party's collective goals.

If this interpretation of Guterbock's argument is correct, we are left with two problems to resolve instead of one. The first is why, before socialization into the party, a sufficient number of Chicagoans were prepared to accept a lower than average income when so few New Yorkers were willing to do so. The second problem is why the socialization process seemed to work in Chicago, so that campaign efficiency was not seriously weakened, whereas in New York most reports in the 1970s indicated that club members, and even officials, were lax in performing their campaign duties. To the latter problem there would seem to be an obvious answer. Once the clubs' memberships had started to decline, and the clubs had become decreasingly important in campaigns, the collective goal of 'team effort' was undermined. That is, the loyal performance of duties was difficult to maintain when many of the party's campaign functions had been transferred to individuals and to other organizations. There is, probably, a vicious circle of party organization decline. However, the former problem is more difficult to resolve, and neither Guterbock's evidence nor that of anyone else does more than suggest possible explanations. Perhaps the most plausible speculation is that the decisive differences concern the interaction of party centralization and governmental decentralization. Chicago's governmental system was decentralized in a very different way from that of New York. New York had boroughs which were a unit of government and the unit on which party power was based. Chicago's government was 'unitary' rather than 'federal', but power was so fragmented among a number of independently elected officials that it would be difficult for any of them to become independent controllers of patronage in the way of New York mayors. But this very fragmentation made it possible for the Chicago machine to be revived by Richard Daley in the mid-1950s. In 'making the city work', Daley followed the example of some mayors elsewhere (such as Stapleton) and appeased the middle classes; he was able to revive an efficient patronage machine after it had declined for some years under Mayor Kelly.

In effecting this revival, Daley benefited from having no rivals with semi-independent power bases in city government; he was also fortunate to be attempting it before the advent of new campaign techniques at the local level. Having re-established party control in the 1950s, Daley was then able to prevent the patronage which emanated from the Great Society programmes from falling into the hands of the party's potential rivals. Thus, what was a catalyst for further machine disintegration elsewhere did not harm the Chicago party. Now, we might suppose, though it is only a supposition, that the effect of continuing centralized control over patronage on recruitment to the patronage jobs would be twofold. First, where there was a regular, organized, city-wide system of employing party people, the jobs would be more attractive to potential party activists, even when they were relatively low-paid ones. The value of the jobs' security would be more evident than when there were many employers of patronage labour operating on a more informal basis. Secondly, party centralization would to some extent allow jobs to be offered where there was a market for them; even though, traditionally, Chicago's ward committeemen had a certain number of jobs at their disposal, jobs which could not be filled in one ward could be filled in another. In New York, on the other hand, a job in Brooklyn which could not be filled in that borough was not usually made available to another borough's party.

Thus we are suggesting that, if New York's 'federal' parties made them particularly susceptible to rapid decline, the party in Chicago was even more unusual in having governmental power so decentralized that a machine could be rebuilt as late as the 1950s. There is no denying that in some conditions hierarchical patronage parties could survive in America. Chicago was not the only example; in very different conditions the Republican machine in Nassau County, New York, survived, and there are a few other examples besides this. Our argument is that it is misleading to see these cases as anything other than exceptional – they are different phenomena from the cases where patronage survived but came to be controlled by individuals with semi-independent power bases. With the latter, patronage was no longer a barrier to penetration of the parties by the forces which were threatening them in the 1960s.

Under Daley the revived Chicago machine survived for twenty years at a time when nearly all other Democratic machines were in continual decline. Only in 1979, with the election of dissident regular politician Jane Byrne, did major signs of strain appear; these were greatly accentuated in 1983 with the election of the city's first black mayor. The success of the regular organization in the 1960s and 1970s in preventing the rise of alternative leaderships, can be seen in the survival of its formal opposition in the Democratic party, the Independent Voters of Illinois (IVI). Unlike the other Democratic club movements of the 1950s, the IVI's membership actually increased in the late 1960s and 1970s – in 1975 its membership was double that of 1964.[33] Protess and Gitelson rightly attribute this success to its continued exclusion from power; for Chicago amateurs there was still an enemy to fight against. However, it should also be pointed out that it could thrive because there were few individual elected politicians around whom alternative organizations and movements could develop. In many ways the IVI was as much an anachronism in the 1970s as the party organization it was trying to overthrow. Naturally, it must be re-emphasized that this account of the difference in the experience of the Chicago and New York machines is somewhat speculative, but it is, perhaps, as much as we can expect given the lack of research by political scientists on activist recruitment in America.

Were it not for three related features of the New York City regular (or professional) party organizations, almost certainly their collapse would have been far more complete than it has been. Two of these features are examined shortly – the complex procedures by which candidates gain access to the party ballot and the partisan manner in which election day practices are controlled. These two features were present in a political world in which the judges of the courts with authority over the misuse of election procedures, and over election day malpractices, were patronage appointees. Court patronage, then, was the third feature of New York politics which ensured a continuing function for the party organizations. Apart from the judges' regulation of election petition disputes and the behaviour of election inspectors, there have been three main elements to judicial patronage in New York.

First, with the exception of Manhattan, the party leaders in the boroughs effectively controlled the party's nominations to both the Civil Court and the Supreme Court. Although even in Brooklyn there were instances in the 1970s of reformers inflicting electoral defeat on the regular organizations' nominees, these low-level contests were especially difficult elections in which to mobilize the electorate against the regulars.[34] Their power to nominate judges provided an important source of income for the regular parties, often amounting to tens of thousands of dollars for each position.[35] This helped to finance the activities of the clubs, including their electoral activities, so that the system was partly self-sustaining, though the other two elements contributed to its sustenance. The second element was that the courts provided employment for clubhouse activists. Most of these positions were those of court clerks, but some of them were of the 'no show' kind. It has been estimated that there were several thousand patronage positions in the New York court system and that, for example, most of the jobs available to Brooklyn boss Meade Esposito were in the borough's court system.[36] Thirdly, the spoils of lucrative legal business at the disposal of judges, especially those in the Surrogates' Courts, were distributed only to approved lawyers – that is, lawyers whom the party organizations designated as appropriate recipients. In Brooklyn fewer than fifty lawyers received the major share of receiverships, guardianships, and refereeships from that borough's courts.[37] Thus, the court system both provided employment for some party workers and also tied in compliant lawyers to the operation of the party organization.

Now there is a misconception about the role of court patronage in New York's parties which, among other places, is to be found in a well-known political almanac. The authors claim: 'It should be apparent to anyone by now that these men [the county party chairmen] have little to do with election outcomes. The machines are in another business now: the brokering of judicial patronage; a county chairman's endorsement is generally a liability rather than an asset.'[38] Three points should be noted about this view of judicial patronage as being a replacement for other electoral activities. The first is that judicial patronage did not replace an older form of patronage politics – it was merely all that remained unmodified from the older style of machine politics. Moreover, although the preservation of what remained of the regular party organizations did depend largely on their retaining control over the courts, this did not mean that they abandoned their older business. In the 1970s the county organizations still benefited from influencing the mayoral election, because of the remaining patronage there, just as whatever control they could muster over their legislative delegations in Albany also enabled them to act as political brokers. The final point is that, just as the incentive for them to participate in non-judicial elections was not removed, so the county organizations did not lose all their influence in this arena. It varied with respect to the levels of office involved and also between boroughs, but it is misleading to conceive the county leaders as being powerless in these elections. Overall, they retained far more influence over the nomination and election process than their counterparts in Denver, or in many cities.

Organization control of the courts is related to a second feature of party politics in New York which helped to limit the decline of the clubs – the procedures for getting on to the party ballot. As we have noted, for a candidate to appear on the ballot in the state, he first had to obtain the signatures of a certain number of registered party voters resident in the area for which he was seeking election, and the number required varied with the office. In theory, this would seem an uncontroversial method for restricting access to the ballot, but in practice it served to bolster the clubs. The reason for this lay in the state's election laws which were filled with technicalities, so that even the most conscientious collector of petitions might submit many which could be shown to be invalid in court. There were two main consequences of this. One was that the collection of signatures on a petition became a difficult task for the political novice. To be sure of appearing on the ballot many more signatures than were actually required had to be gathered, and knowledge of both electoral law and an area's residents provided any group with a distinct advantage. It was precisely this kind of knowledge that, traditionally, the New York clubs possessed, and it is not surprising that, whenever possible, most candidates continued to rely on clubs to make a contribution to this aspect of their campaigns. Regular and reform clubs performed these tasks and, in delegating one specialized and labour-intensive function to these organizations, the candidate could focus his own campaign organization on other electoral activities. Thus, in the 1970s, party activists were still playing a role which their counterparts neither in Denver and the East Bay, nor most other urban areas, now had. The second consequence of the legal technicalities was that, in at least some primaries in any election year, there would be a court challenge by one of the contestants to the validity of his opponent's petitions. Moreover, these challenges were even more common than they would have been because state law permitted the cross-endorsement of candidates. Hence, a Democrat and a Republican might be rivals for the Liberal party nomination and challenge each other's petitions for designation on that party's ballot line.[39] Of course, the judges hearing these cases were elected on a partisan ballot and, since all but a few of them had close ties to the Democratic party, especially its regular organizations, their decisions normally helped to perpetuate these organizations.

Their perpetuation has been facilitated further by partisan involvement in the regulation of activities in the polling stations. On election days the city usually employed about 23,000 inspectors to supervise the stations, but these appointments have been strictly partisan ones. As patronage, their importance has not been great but, in many election districts, having their own supporters as inspectors has been reckoned to be worth at least a few votes for the candidates. In some instances the result of the election itself has been affected by the inspectoring system. (The most celebrated case in the period 1960–80 was the Rooney–Lowenstein primary in Brooklyn in 1972, after which a judge ordered a second election because of irregularities in the polling booths.)[40] In those parts of New York still controlled by the regular organizations it was their clubs which supplied the inspectors, with reform Democrats selecting these officials only in the districts they controlled. For the regular clubs, the inspectorate system provided an important source of influence *vis-à-vis* public office-seekers, because it created an additional barrier in the election process should an office-seeker fail to reach agreement with the regulars. For the reform clubs, it constituted a similar political resource, though arguably a less valuable one, since their political ethos usually restricted their willingness to resort to electoral malpractices. However, this difference between the two types of club should not be exaggerated: if they were less likely to commit overt election fraud than the regulars, knowingly they allowed it to occur sometimes. This was one of the controversies in the well-known dispute over the endorsement of John Lindsay by the Village Independent Democrats (VID) in 1965. The VID leaders endorsed the Republican mayoral candidate shortly before the general election. At a subsequent meeting of the Manhattan county party there was a motion to censure them in their capacity as district leaders. One of the main charges laid against them was that they had not assigned inspectors or poll watchers to polling stations, and that thereby they had left 'Democratic candidates unprotected'.[41] The implication of those who laid the charges was that knowingly the reformers had allowed others, presumably Republicans and Liberals, to affect the vote total in their favour because the credentials of all who came to vote had not been checked.

Yet, if there are factors which have limited the decline of clubs which recruit professional activists in New York, it cannot be denied that a major decline has occurred. In 1932, when Democratic reform clubs were unknown, Peel's study revealed that there were 700 Democratic clubs in the city.[42] When Adler and Blank replicated his study in 1972 they found only 110 active regular clubs (and forty-six reform clubs) attached to the Democratic party.[43] They calculated that the membership of the regular clubs was approximately 59,000, although they acknowledged that the clubs exaggerated the size of their membership and that they could not estimate what proportion were active members. Peel had not examined club size in 1932, but there is no evidence whatsoever to suggest that clubs had merged or that the typical club was much larger than fifty years earlier. On the contrary, Adler and Blank 'believed that clubs are smaller today' than they were in 1932.[44]

The circumstantial evidence of a dramatic reduction in the total number of active regular club members is overwhelming. Typical of such evidence is one experience of the Beame campaign in the Bronx in the 1977 mayoral election. Mayor Beame was supported by most of the regular organizations, and the Bronx county party was one which had remained largely unpenetrated by reformers, despite the electoral success of non-regular candidates in the borough. However, Beame had to pay $10,000 to a non-party organization to obtain a 'prime voter' list – precisely the kind of information which in the past Democratic captains were supposed to have readily available.[45] The Bronx party simply did not have a sufficiently comprehensive or active grassroots organization to provide even basic electoral information. At the same time, it was also reported that elections for district leaders were generating little concern among either voters or politicians, and that the apathy displayed contrasted with the intense competition surrounding these offices even ten years earlier. Certainly, the evidence from the early 1960s, including Edward Costikyan's published account, indicates great electoral activity by the clubs.[46] Nevertheless, to recognize that the collapse of interest in these offices has happened comparatively recently is not to suggest that the decline in club memberships revealed by Adler and Blank occurred wholly, or even mainly, within this period. Fragmentary and circumstantial evidence indicates that the decline in the party work-force from the regular clubs has been continuous for at least thirty years, though the rate of decline may have increased from 1960 onwards. Despite this decline, competition for party offices remained intense so long as the party remained the most powerful influence in the nomination and election process. The erosion of this power, beginning most clearly with the Wagner re-election in 1961 and continuing from then on, made party offices less attractive, and by the mid-1970s competition for party posts had collapsed in many districts. If patronage was not dead in New York politics, the hierarchical party organization was expiring.

## 3. The Disappearance of the Amateur Activist

After the surge of interest in the Democratic party generated by the 1952 Stevenson campaign, purposive participation in the party did not develop uniformly throughout the country. In California the CDC, the most successful of the club movements studied by James Q. Wilson, was already in decline by the end of 1963. At the same time in Denver, the liberal activists were consolidating the advances they had made the previous decade, which had culminated in control of county and state party organizations. In some cities the amateur movement continued to expand for two or three years. New York is an example of this. Reformers in Manhattan were now in a position effectively to cripple the operation of a regular organization at the county level, while a major reform movement was developing in the Bronx and there were smaller expansions in Brooklyn and in Queens. Generally, however, circumstances were not that favourable to the growth of amateur politics between 1963 and 1965. Social changes were beginning to produce increased competition for the time of the solidary participant, and there were few stimuli for purposive participation in Democratic parties. The Democrats were in power in Washington and in many of the states, and the main issue of the period, civil rights, was not one which was likely to encourage participation in a party by the white middle-class activists – there were more appropriate vehicles for that.

In 1965 the world of the amateur Democrat changed dramatically with Lyndon Johnson's commitment of American troops to a combat role in Vietnam. For more than three years the war pitted Democrat against Democrat and, through its effects on party participation, accelerated the process of party collapse in America. If the divisiveness of the war stimulated mass activist mobilization, it weakened in two ways the amateur Democrat movement as a primarily party-oriented movement. First, intensity of feeling about the war by its opponents reduced activist loyalty to the party; for many of the older participants, as well as for most of the newly mobilized, party became a mere device for promoting an issue. Secondly, the division caused by it being a Democratic President's war fostered activism outside the party – in the campaigns of individual candidates, in interest groups, and in direct action. This removal of the party as the principal focus of purposive participation was to affect the role the party could play in the later development of issues relating to the environment, 'good government', and consumer affairs, even after Johnson's war had become Nixon's war. However, it must be stressed that the effect of the war was not always the immediate destruction of the Democratic party as the main channel for purposive participation at the local level. In some cases, as with the CDC in California, this did happen. But in cities such as New York, where the amateur movement had a strong orientation to local issues which had not been resolved and a permanent institutional structure at the local level, the war arrested the growth of the movement but did not eliminate it. In New York the history of both amateur and professional clubs after 1965 was to be similar – they would decline together. Furthermore, in some instances, where the party was both open to the newly mobilized and appeared still to be a major electoral intermediary, it became a vehicle much used by purposive activists. Denver was an example of this, though the era of mobilization through the party lasted only about seven years. As we have argued elsewhere, an 'exit' to other types of political organization was emerging in Denver and, as we shall see in Chapter 5, the mass mobilization by amateurs occurred at the same time as the power of the county party organization was waning rapidly.[47]

Having outlined the impact of the war very generally, we must now turn to examine each of the three areas in more detail. We begin with the East Bay where, at least superficially, the amateur movement had enjoyed most success in the 1950s. There is widespread agreement among its then leaders that, in respect of the numbers of people involved and their contributions of time to electoral tasks, the amateur Democrat movement in the East Bay reached its peak in the mid-1950s.[48] After that, membership and campaign contributions declined, although with success in state and local elections, most notably in 1958 and 1961, the clubs, the Berkeley Caucus, the 18th ADDPO, and the 7th CD organization remained effective until 1966. We consider the fate of the party organizations in the next chapter; for the moment our concern is with the activists and with the informal party structures – the clubs and the Berkeley Caucus. However, we cannot understand fully what happened to these bodies after 1966 without considering the relationship between the clubs and the party structures at the state level.

In the last chapter it was argued that Wilson's analysis of the club movement in southern California is not applicable to the East Bay. However, he was correct in arguing that there was a tension between the relatively weak state party organization and the CDC.[49] The rise of the CDC had provided the former with a powerful rival, but the tensions within the Democratic party did not become serious until Democrats wielded power in Sacramento. Before 1959, at the very least, the CDC could claim credit as a body which assisted in the election of Democrats in the state. By making public endorsements before the party primaries, the CDC had helped to reduce the advantage of Republican incumbents running in Democratic primaries, because one Democratic rival would be recognized publicly as the party's semi-official candidate. The CDC had further attempted to reduce the Republican advantage by requiring that any candidate who sought the CDC endorsement must agree to withdraw from the primary if he did not obtain it. Particularly in southern California, the CDC itself had been a major power in the nomination process and the members of the affiliated clubs provided manpower which the party had always lacked. Expanding into areas where the initial Stevenson movement had been weak, the CDC continued to gain recruits even while in places such as the East Bay the liberal activist movement was passing its prime. In the year of its formations, 1953, there were 20,000 members in 100 clubs; by 1963, when state-wide affiliation was at its peak, there were 75,000 members in 650 affiliated clubs.[50]

However, in 1959 the cross-filing primary was abolished. The next few years were ones of conflict between the CDC and the Democratic Governor, state legislature leadership, and the formal party organizations. The public policies advocated by the CDC were far more liberal than those of the elected public officials, who came to see the CDC as a potential liability for them at future elections. The most publicized of these disputes was with Jesse Unruh, the speaker of the state Assembly. It began in 1963 and continued until it was overtaken by the Vietnam issue in 1965. Then, when the new CDC president vehemently criticized Lyndon Johnson, and the CDC called for the immediate withdrawal of American troops, Governor Pat Brown was drawn into the dispute, and the CDC's break with the party leadership was complete. From 1965 it became an organization of relatively minor importance in state politics, having a membership by the early 1970s one-seventh the size it had in 1963. There can be little doubt that the war issue facilitated its rapid demise; the more interesting problem is whether it would have declined in the absence of a divisive foreign policy under a Democratic administration.

The strongest argument which can be made supporting the claim that the CDC collapse would not have occurred is that periods when the party controls major public offices pose problems for amateur organizations. In this view, the conflicts with Unruh and others were more easily resolvable with Ronald Reagan as Governor than under a Democratic monopoly of power in Sacramento. Once Democrats were out of office co-operation would have replaced conflict. Yet, even if we accept this argument, there are two reasons for believing that CDC decline would have continued after 1966. First, the very conditions which had encouraged its growth – the cross-filing primary and a Republican monopoly of power in the state – had been removed. There was no possibility of cross-filing being revived and, in spite of Pat Brown's defeat in 1966, the Democrats were in a far more competitive position in the state than they had been in the early 1950s. The Republicans were no longer an entrenched party of government. Secondly, those whom the CDC had helped put in office could now take increasing advantage of new campaign techniques, including television, which had been pioneered in the state. It is important to recognize how different the political world was in 1965 compared with twelve years earlier when the CDC was formed. Only a minority of homes possessed televisions in the early 1950s, but televisions were owned by most households in the mid-1960s. Moreover, although Richard Nixon had used the medium in a masterly way in 1952 to save his career with the 'Checkers' speech, political advertising by candidates was still in its infancy. By 1965 there had already been two 'television age' presidential campaigns, and a number of gubernatorial and senatorial candidates in large states had started to exploit television as a major campaign device. In brief, established politicians in California had fewer electoral resources to lose from a split with the CDC in 1965 than they would have had in the mid-1950s.

With more local orientation than the southern California clubs, the clubs in Berkeley were less affected by the CDC's state level disputes. From 1961 the city was governed by liberal Democrats who were closely associated with the clubs or their political allies the split between elected public officials and activists found at the state level was not repeated here. However, the effect of the Vietnam war on the club movement was at least as devastating as elsewhere, because it followed the radicalization in 1964 of a significant minority of students over the 'Free Speech Movement'. Thus, it was not surprising that in 1965 Berkeley was one of the first campuses to experience large-scale opposition to the war. With easy candidate access to the primary ballot in California, it then became one of the first congressional districts in which a specifically anti-war candidate appeared against an incumbent. The incumbent, Jeffrey Cohelan, was one of the most liberal Democrats in the House and one of the main beneficiaries of the success of club politics in the city.For failing to criticize Johnson's war vigourously, Cohelan was challenged by a student opponent in the 1966 Democratic primary. He won that election but with only 55 per cent of the vote, and he actually lost in the Berkeley part of the district. This was the first major intrusion of the previously campus-centred radicalism into city and Democratic party politics. Hence, just at the time that the amateur Democrats had consolidated their position *vis-à-vis* organized Republicanism in Berkeley, so their movement was split by the war and faced a newly mobilizing political force. It was a force with beliefs and methods which were only partly compatible with those of the amateur Democrats, and what ultimately destroyed this earlier movement was the shift by the radicals into community politics. From being centred on the University of California in 1964, radicalism moved on, via the 1966 campaign, to local government. The interplay of their direct action tactics and police violence brought about conflict with the new liberal establishment. Thus the war was a catalyst for the breakdown of party politics in local government, and it was this which destroyed the club system. By the early 1970s there was only one Democratic club left in the city. The clubs could never have channelled the new participation, and they would have remained divided over the war until Cohelan was replaced, but they might still have survived had they not had to face the radical challenge at the very heart of their own activity.

In 1967 the radicals selected four candidates to run as a slate in the city council elections. This broke the now established pattern of party competition in city elections, of a Democratic slate versus a Republican slate. The Berkeley Caucus was divided between its more conservative club members and those sympathetic to the New Left who favoured endorsing some radical candidates. Because it was able to avoid a split the Berkeley Caucus remained intact after 1967, and liberal Democrats continued to control the city council. However, the Democratic party was rapidly losing its role as the co-ordinator of non-conservative political activism in the area, for many of the new radicals had either no commitment to the Democratic party or were openly hostile to it. After 1966 the 18th ADDPO ceased to exist, and the effective end for the Berkeley Caucus came in 1969. That year it was approached by the radicals with a view to the Caucus selecting one of them as a member of the four-person election slate. On a vote the Caucus rejected the radical candidate, and the radicals subsequently ran a two-person slate of their own. Neither of these candidates won, but they took sufficient votes away from the Caucus slate for a Republican and an Independent to be elected. Democrats no longer controlled the city council and, as we shall see in Chapter 5, organizations related to the Democratic party were no longer major participants in Berkeley city elections. In Berkeley city elections. In Berkeley party politics at the local level was to be replaced by neo-party politics.

The demise of the amateur Democrat movement in the East Bay occurred in circumstances which were not replicated in New York. The legally-defined parties were extremely weak; the liberals were in power in the city that was the centre of their movement; and that city was relatively small and dominated by a university, so that radicalism in the latter could have a major impact on the government of the city and on the local Democratic party. Yet, although they did not disappear like the clubs in the East Bay, the amateur clubs in New York were much weaker by the early 1970s than they had been in 1965. To understand this decline it is first necessary to distinguish between three kinds of club. The first kind are 'revolt' clubs – regular clubs which seek to defeat other regular clubs which control their districts. They used to be the main form of opposition in the Democratic party, and Peel's study discovered 120 of them in New York in 1932.[51] The second kind are 'old' amateur clubs; they developed between the late 1940s and the mid-1960s and were the subject of Wilson's study. The third kind are Democratic clubs or groups which developed in opposition to the Vietnam war. At the end of 1968 these 'new' amateurs sought an amalgamation with the 'old' amateurs and an 'umbrella' organization, the New Democratic Coalition (NDC) was set up in 1969 to achieve this. In many ways the NDC was a replacement for the Committee of Democratic Voters, to which most of the 'old' amateur clubs were linked in the 1950s. However, not all the 'old' amateur clubs joined the NDC: the Bronx clubs were, perhaps, the most important non-participants. The distinction between the three types of club is significant when we turn to consider how membership in the amateur movement has declined, for there are no commonly agreed statistics on membership, and the estimates available refer to different population bases.

In 1961 Wilson estimated that there were 8,500 members of amateur clubs in Manhattan, but he provided no estimate of the size of the movement in other boroughs.[52] They controlled no Assembly Districts outside Manhattan, and it is clear that Wilson believed that club membership there was small. Between 1961 and 1968 the amateur clubs did make inroads into the electoral monopoly of the regulars in the Bronx, and also in small areas of Brooklyn and Queens, but no statistics are available relating to membership growth in these boroughs. In their study of the early 1970s, Adler and Blank considered the 'old' and 'new' amateurs together, though the way they presented their data was confusing as it appeared to incorporate 'revolt' clubs as well. They estimated that there were over 12,000 members in these clubs, but they admitted that, like the regulars, the amateur clubs may have exaggerated their size when responding to the questionnaire. Comparing the Adler and Blank evidence with that of Wilson would suggest that membership in amateur clubs increased during the decade between the two studies, and there are two reasons for believing such an increase might have occurred. The reform movement did spread outside Manhattan in the early 1960s so that, unless there was a great decline in Manhattan at the same time, total membership would have increased in the city by 1965 or 1966. Even a subsequent fall in membership might well not have pushed the club movement back to its 1961 position. Again, there are the claims of the NDC to have had a state-wide membership in its early years of 20,000; most of these members were residents of New York City.[53] But the fragmentary evidence we have about *active* membership suggests a very different situation – that there was a much smaller active core in the early 1970s than ten years before. For example, even as early as 1971 the *New York Times* was reporting the decline of both professional and amateur clubs in the city. As an instance of amateur decline, it cited the Murray Hill club in Manhattan which had only 125 members, of whom about one-fifth would attend meetings; this compared badly with membership turnouts of more than 200 people in 1961.[54]

Like other 'old' amateur clubs in New York, Murray Hill was adversely affected by two developments between 1965 and 1973. The first was that, with John Lindsay as mayor, much of the immediate relevance of the amateur Democrat movement to potential activists disappeared. On the one hand, unlike his predecessor Wagner in 1961, Lindsay did not mobilize these elements directly. Instead, it was an older organization of liberals, the Liberal party itself, which Lindsay used in 1965 and this, together with the John V. Lindsay Associations, was his channel of patronage when he assumed office. On the other hand, it became less easy for reform Democrats to dramatize city politics as an 'us–them' conflict which could be used to recruit members. The Lindsay mayoralty blurred these divisions, just as Wagner in his 1961 re-election had highlighted them. Whereas in Chicago continued exclusion from power helped to increase reform club membership during the period 1966 to 1973, in New York Democratic reformers twice helped to elect Lindsay without either obtaining power for themselves or being obviously excluded from it.[55] They lacked the important resource for mobilization of being an entrenched opposition. Furthermore, it was not just Lindsay's political base or his party labels which led to the de-emphasizing of the amateur–professional conflict during his administration. The two main issues to emerge during his years of office, the growth of municipal union bargaining power and race relations, cut across this line of cleavage.

The second development was that the emergence of the war issue, and later environmental and consumer issues, served to undermine the balance between local and national concerns which had been established by the 'old' amateur clubs. For example, the amateur clubs in Chicago and the East Bay, like those in New York, had endured beyond the years of the Stevenson presidential candidacies because they directed the activity of those initially mobilized by national political issues into organizations with both a national and a local focus. For the clubs, the Vietnam war was a serious threat, not simply because of its divisiveness, but because it drew attention to the limitations of Democratic clubs as organizations for mobilizing on a single issue. The clubs had other concerns besides the war, and specialist single-issue groups were an obvious alternative to them. Only briefly in 1968 and shortly after, at the time of the founding of the NDC, did anti-war activism probably lead to an increase in the number of participants in the New York club movement. Furthermore, it did so through the creation of new clubs rather than the re-invigoration of old ones. Much of the membership increase was short-lived, and certainly it had only a limited impact on the continuing struggle between regulars and reformers. While it could claim 20,000 members shortly after its formation, by 1979 the NDC had to acknowledge that state-wide there were a mere fifty-five affiliated clubs with a total membership of only about 8,000.[56]

The amalgamation of the 'old' and 'new' amateur movements was not always achieved without conflict. In particular, in the 65th and 67th Assembly Districts on the West Side of Manhattan the conflict was a bitter one. In 1969 younger, ex-McCarthyite activists defeated several older reformers in elections for District Leaders, and they even succeeded in electing one of their members as a state assemblyman.[57] Their principal criticism of the 'old' amateurs did not concern the war, but involved a local issue – the alleged failure of the 'old' amateurs to give sufficient attention to their districts. The rivalry between the two groups was intense for a couple of years, and temporary alliances with regulars were employed as a means of weakening their opponents. But the conditions in which this occurred were very different from those in Berkeley. There was no question of the dispute suddenly destroying the older amateur movement city-wide nor that a neo-party system might develop in local government. The state of politics in the rest of the city necessitated that a reconciliation be brought about, and eventually it was.

Finally, we must turn to Denver, a city in which in the short term the Vietnam war had a somewhat different impact on the Democratic party, but here again in the long term it did not help recruitment to the party. In the same way as many other cities, Denver had experienced an upsurge of middle-class activity in the Democratic party in the 1950s, which in this case led to outright control of the county party by liberals in 1962. Not surprisingly, the movement was centred on the newer and more affluent neighbourhoods in the south and east of the city. Throughout the 1950s population shifts in the city had tilted the balance on the party's Executive Committee away from the conservatives in the older, 'ethnic' neighbourhoods. Unlike the amateurs in both California and New York, the liberals in Denver did not organize in clubs separate from the formal party organizations, and it is useful to compare the Denver experience with that of the other two areas. In contrast to California, the Denver party organizations in the 1950s were still crucial in determining nominations for public office and in campaign activity. Given that these structures were open to activists, there would have been no advantage in replicating them. Once there were enough liberals to undertake the 'everyday' tasks of party work, it was relatively easy to participate in the party organization and subsequently control it. In New York clubs were used by liberal activists because it was only through prior club organizations that access to the formal party organizations could ever be attained. There the positions of Assembly District committee member and leader were open to serious challenge only by copying the regulars – organizing in an extra-legal body, a club, and using its resources to compete for the legally constituted positions. Although Denver had a relatively strong party organization, there were several reasons why entry to the party had been easier than in New York – at least since the 1920s. First, the political values dominant in the state restricted the potential for using public affairs for private gain; in Elazar's language, the state had a moralistic political culture.[58] Secondly, the partial separation of Stapleton's machine from the Democratic party meant that an open party was not a direct threat to the patronage operation, and thus efforts to limit entry were not necessary. Thirdly, any such efforts might have succeeded only with great difficulty. The liberal activists were concentrated initially in the newer housing developments or the more Republican parts of the city where Democratic party organization had been weak. Moreover, it was more difficult to stop people attending caucus meetings in Denver, which was the first stage of nomination to party office, than it was to invalidate election petitions in New York, the first stage of the process there and one which advantaged the already mobilized. Thus, the very openness of the Denver party's caucuses and organization led to it being used as a device for opposing the Vietnam war, and that experience produced high levels of participation in the party in the early 1970s when other issues had become more salient. The war's impact was different in Denver than in New York or the East Bay; as we have argued elsewhere, it stimulated activism within the party, rather than generating conflict between activists.[59]

After the first few years of the 1970s, however, the experience of Denver came more closely to resemble that of Democratic parties elsewhere; indeed, even by 1971 concern about single issues was leading some activists away from party structures, and towards other forms of organization. One of the main instigators of this was the insurgent liberal who defeated Congressman Byron Rogers in the 1970 Democratic primary. He revived the use of the initiative referendum in Colorado, and thereby popularized non-party campaigning among liberal activists. This device was of little value on the issue of the war, which by then was becoming 'Vietnamized', but it could be, and was, used on consumer and 'good government' causes which were becoming prominent in the state. Among a number of referendums which liberals placed on the ballot in 1972, the most famous was the one to stop funding for the proposed 1976 winter Olympics in Colorado. Undoubtedly, the ease with which activists could be mobilized around these causes was partly the result of heightened political activity over the war issue. The war was a catalyst because it had shown that the electoral process could be used to defeat those in power; but the activists had also learnt that they could not expect electoral success to bring immediate policy changes or changes in attitude among older party élites. 'Exit' from party-centred activity followed.

By the late 1970s the Denver party was a predominantly amateur party with members who were far less active in the party than their predecessors had been. The amateur orientation of the organization members was revealed in their responses to a survey question on their reasons for continued participation in the party. Asked to assess the importance of the following reason, 'I am concerned about the policies of national and state governments, and working in the Democratic party gives people like me some influence over what governments do', 94 per cent of the Executive Committee and 88 per cent of the precinct committee members claimed that for them it was either 'very important' or 'fairly important'. Indeed, 48 per cent of the former group (n = 51) and 51 per cent of the latter group (n = 196) cited it as a 'very important' reason. That in the long term the relatively peaceful transition the party experienced in the 1960s did not produce a party which was both amateur *and* effective can be attributed to two factors: the incentive to 'exit' to other kinds of political organization and the decline in the party organization's base of power in the late 1960s. (We examine this second factor in Chapter 5.) One consequence of these two developments was that recruitment to the county party became difficult though, given the high level of population mobility in a city such as Denver, the vacancy rate for precinct committee positions might not appear to be high. However, a vacancy rate of 18 per cent in 1979 must be seen in the context of the inexperience of those who were serving as precinct committee members. Thirty-six per cent of the members had served in their posts for less than fifteen months, and only 8 per cent had served for more than ten years (n = 199). No less than 47 per cent of all committee posts were either unfilled or filled by recruits of less than fifteen months standing. High personnel turnover of this kind would suggest that the organization might face problems in getting campaign tasks performed, and this was the case. The party had great difficulty in obtaining the amount of campaign work which earlier precinct workers had undertaken. Many of those who served on the party's central committee in the 1970s did so because there was no one else who wanted the job; this meant that there was no incentive for district captains to remove inactive members.[60] Often, in recruiting people to run for the committee, the captains simply had to hope that those who had expressed an interest in political issues might carry out some campaign tasks. Naturally, most captains and party leaders preferred the public to believe that there was an 'army of party troops', which could be pointed to on paper; they had no desire for a purge which would encourage those who did little work for the party to vacate their positions. Estimating the size of the 'paper army' of supposed party workers is not easy. Few of the committee members actually admitted that they did not usually take part in election campaigns.[61] Yet, virtually all the party leaders denied that the amount of work they did was significant; the most commonly cited claim was that no more than one-quarter of the members performed more than a few perfunctory tasks each election year, either for the party or individual candidates.

If the Denver party organization officials, candidates, and public officials had a rather low opinion of the quality of the organization membership, then it must be admitted that by 1979 the members themselves accepted the secondary role that the organization was playing. Asked to comment on the statement, 'In my experience, the campaign organization of a Democratic candidate is usually much more important than the Denver Democratic party in helping to win the general election', only 3 per cent of the central committee claimed to 'strongly disagree' and only an additional 9 per cent 'mildly disagreed'.[62] Moreover, the committee members also gave the Democratic party a low rating in relation to other actors, as a source of campaign funds to Democratic candidates. Asked to comment on the importance of the contribution of six sources of funds (the candidate himself, his friends and contacts, the Democratic party, labour unions, other interest groups, and fund-raising efforts by the campaign organization), the committee members finding the party either 'very important' or 'fairly important' did constitute 64 per cent of the respondents, compared with 23 per cent who found it 'fairly unimportant' or 'completely unimportant' (n = 204). However, only the candidate himself was rated lower than this as a contributor, and the party fared poorly in comparison with the campaign organization which was regarded as important by 90 per cent of the respondents.[63] That the organization members should not have displayed loyalty to it in their responses, by possibly over-valuing its importance, is not surprising because so few of them had given the long service which might breed loyalty. In many ways, therefore, the Denver party had become the opposite of the Chicago party in the 1970s – there was no organizational ethos binding the members in the pursuit of collective goals.

Despite its relatively easy adjustment to the political disruptions of the mid and late 1960s, the Denver party became one which could not maintain an effective work-force. A high turnover rate among organization members, a failure to carry out basic electoral tasks, and a view of the party as less important than other actors in the electoral process were the main features of this new-style party organization. This might seem surprising when we remember that Colorado had institutional arrangements which were conducive to fairly strong parties, and had actually practised strong party politics until the mid-1960s. In seeking to explain this rapid collapse, we must turn to factors other than the recruitment of activists, for this was only one of the elements in the demise of the party.

## Notes

[1] Peter B. Clark and James Q. Wilson, 'Incentive Systems: A Theory of Organization', *Administrative Science Quarterly*, 6 (1961), 129–66, this classification is employed in Wilson's empirical study, *The Amateur Democrat*.

[2] James Q. Wilson, *Political Organizations* (New York: Basic Books, 1973), pp. 33–4. It is important to note that Wilson himself uses the term 'goal' in a much more restricted way than we are using it here, and talks about the *incentives* as being material, solidary or purposive. But, obviously, someone 'responding' to, say, a material incentive can be said to have a material goal (or objective – at least in the 'everyday' sense in which these terms are used.

[3] The anti-hanging lobby is an example used by Wilson in discussing the distinction between types of incentive; *Political Organization*, p. 35.

[4] Wilson, *The Amateur Democrat*, pp. 2–11.

[5] This point is made by Adler and Blank who point to a term coined by Goetchus – 'hairshirt' reformers – to refer to the kind of activists Wilson sought to isolate as 'amateurs'; Norman M. Adler and Blanche D. Blank, *Political Clubs in New York* (New York: Praeger, 1975), p. 175. (V. Goetchus used this term in an unpublished manuscript, 'The Village Independent Democrats', written in 1963.) That the issue-oriented activist movement of the 1950s was more broad based than Wilson suggests is also argued in Ware, 'Why Amateur Party Politics has Withered Away: The Club Movement, Party Reform and the Decline of American Party Organizations'.

[6] C. Richard Hofstetter, 'The Amateur Politician: A Problem in Construct Validation', *Midwest Journal of Political Science*, 15 (1971), 36.

[7] Jeffrey L. Pressman and Dennis G. Sullivan, 'Convention Reform and Conventional Wisdom: An Empirical Assessment of Democratic Party Reforms', *Political Science Quarterly*, 89 (1974), 557–8.

[8] Hofstetter, 'The Amateur Politician', p. 36.

[9] Wildavsky, 'The Goldwater Phenomenon: Purists, Politicians and the Two-Party System'.

[10] Other uses include that by Keech and Matthews, who employ the term to denote those who are intermittently active in a party; William R. Keech and

Donald R. Matthews, *The Party's Choice* (Washington, DC: Brookings, 1976), p. 159.

[11] See especially Bruce M. Stave, *The New Deal and the Last Hurrah* (Pittsburgh: University of Pittsburgh Press, 1970); Raymond E. Wolfinger, 'Why Political Machines have not Withered Away and Other Revisionist Thoughts', *Journal of Politics*, 34 (1972), 365–98; and Michael Pinto-Duschinsky, 'Theories of Corruption in American Politics', paper presented at the Annual Meeting of the American Political Science Association, Chicago, 1976.

[12] Raymond E. Wolfinger, 'The Development and Persistence of Ethnic Voting', *American Political Science Review*, 59 (1965), 896–908; Michael Parenti, 'Ethnic Politics and the Persistence of Ethnic Indentification', *American Political Science Review*, 61 (1967), 717–26 Michael Novak, *The Rise of the Unmeltable Ethnics* (New York: Macmillan, 1971).

[13] Verba and Nie reported that 8 per cent of their survey claimed membership in 'Political groups, such as Democratic or Republican clubs, and political action groups, such as voter's leagues', but only just over 5 per cent claimed that they were active members. Sidney Verba and Norman H. Nie, *Participation in America* (New York: Harper and Row, 1972), p. 42.

[14] Albert O. Hirschman, *Shifting Involvements* (Oxford: Martin Robertson, 1982).

[15] Alan Abramowitz, John McGlennon and Ronald Rapoport, 'The Party Isn't Over: Incentives for Activism in the 1980 Presidential Nominating Campaign', *Journal of Politics*, 45 (1983), 1006–15. For an example of an analysis which views the Democratic party as the victim of anti-partyism, see Ranney, *Curing the Mischiefs of Faction*.

[16] Jack Dennis, 'Trends in Public Support for the American Party System', *British Journal of Political Science*, 5 (1975), 187–230.

[17] *Denver Post*, 5 Nov. 1962.

[18] John J. Harrigan, *Political Change in the Metropolis* (Boston: Little, Brown, 1976), p. 93.

[19] Interview no. 84.

[20] Thomas S. Guterbock, *Machine Politics in Transition* (Chicago: University of Chicago Press, 1980), p. 13.

[21] For example, see Harrigan, *Political Change in the Metropolis*, p. 94.

[22] Wolfinger, 'Why Political Machines have not Withered Away'; see also *New York Times*, 19 Jan. 1975.

[23] Some of the examples mentioned here are cited in *New York Times*, 19 and 20 Jan. 1975.

[24] In the list of political clubs they found in New York City, Adler and Blank identify only one Liberal Club; *Political Clubs in New York*, p. 257.

[25] *New York Times*, 7 Mar. 1971 and 27 Nov. 1971.

[26] *New York Times*, 15 Apr. 1974 and 27 Nov. 1971. As the independent power of the state legislators increased, so country party leaders had more difficulty in getting them to fund 'no show' jobs. In the early 1970s this process was more advanced in Queens than in Brooklyn. Queens leader Troy had to resort to a public appeal to the legislators in 1972 in an attempt to get them to pay the party half the staff allowance.

[27] *New York Times*, 17 Feb. 1979. While the decay in the party's electoral control was more advanced in the Bronx than in Brooklyn, the party organization of the former remained a more exclusively regular one. In 1976 three-quarters of the Bronx district leaders were patronage job holders; *New York Times*, 12 Feb. 1976.

[28] For example, even Guterbock, whose detailed study of the Chicago party lasted several years, acknowledged: 'Since the system operates in secret, the number of patronage jobs and the exact procedures by which they are dispensed are difficult to determine', *Machine Politics in Transition*, p. 15.

[29] Guterbock, *Machine Politics in Transition*, p. 15.

[30] Ibid., pp. 15 and 37.

[31] Ibid., p. 37.

[32] Ibid., pp. 37 ff. and pp. 214–16.

[33] David L. Protess and Alan R. Gitelson, 'Political Stability, Reform Clubs, and the Amateur Democrat', in William J. Crotty (ed.), *The Party Symbol* (San Francisco: W. H. Freeman, 1980), p. 93.

[34] Between 1975 and 1979 nine insurgents defeated regulars for Civil Court judgeships in Brooklyn. *New York Times*, 17 Feb. 1979.

[35] Estimating the amount paid to a party for a judgeship is extremely difficult; however, an attempt was made to do this in 1971. See Martin Tolchin and Susan Tolchin, *To the Victor* (New York: Random House, 1971), p. 146.

[36] *Village Voice*, 23 May 1979 and *New York Times*, 17 Feb. 1979. In 1974 nine of the forty-four district leaders in Brooklyn had court jobs; *New York Times*, 15 Apr. 1974.

[37] *Village Voice*, 3 Apr. 1978.

[38] Michael Barone and Grant Ujifusa, *The Almanac of American Politics 1982* (Washington, DC: Barone, 1982), p. 723.

[39] For an example of this, see the case discussed in the *New York Times*, 8 Aug. 1977.

[40] For a discussion of both this particular case and the more general problem of the inspectorate system, see Walter P. Loughlin, 'Election Administration in New York City: Pruning the Political Thicket', *Yale Law Journal*, 84 (1974), 61–85.

[41] *New York Times*, 19 Nov. 1965.

[42] Roy V. Peel, *The Political Clubs of New York City* (New York: Putnam, 1935).

[43] Adler and Blank, *Political Clubs in New York*, p. 182.

[44] Ibid., p. 36.

[45] *New York Times*, 22 Aug. 1977.

[46] Costikyan, *Behind Closed Doors*. Fourteen years later he claimed that 'Before it was published, however, it became a history of a bygone era', Edward N. Costikyan, *How to Win Votes* (New York: Harcourt, Brace, 1980), p. 6.

[47] Ware, *The Logic of Party Democracy*, Ch. 6 and 7.

[48] Interview nos. 57, 58, 59, and 64.

[49] Wilson, *The Amateur Democrat*, Ch. 4.

[50] Larry N. Gerston, 'How Politicians Choked Out the Grass Roots Movement', *California Journal*, June 1976.

[51] Adler and Blank, *Political Clubs in New York*, p. 177.

[52] Wilson, *The Amateur Democrat*, pp. 34–5.

[53] *New York Times*, 2 July 1979.

[54] *New York Times*, 27 Nov. 1971.

[55] On Chicago see Protess and Gitelson, 'Political Stability, Reform Clubs and the Amateur Democrat'.

[56] *New York Times*, 2 July 1979.

[57] *New York Times*, 19 June 1969 and 13 July 1971.

[58] Elazar, *American Federalism*, ch. 4.

[59] For an account of this see Ware, *The Logic of Party Democracy*, ch. 6 and 7, and Ware, 'Why Amateur Politics has Withered Away'.

[60] Less than one in four of the respondents to the questionnaire 'strongly agreed' with the following statement: 'If I wanted to give up my position within the formal Democratic party organization, it would be easy for a successor to me to be found' (n = 204).

[61] Only 3 per cent of the precinct committee people surveyed admitted that they did not usually take part in any campaign, (n = 199).

[62] The responses to this question were as follows: 'strongly agree' 39 per cent, 'mildly agree' 37 per cent, 'undecided' 9 per cent, 'mildly disagree' 9 per cent, and 'strongly disagree' 3 per cent (n = 204).

[63] The question asked was, 'In your experience, how important have each of the following sources been in raising funds for a Democratic candidate who is contesting a partisan office such as the US Congress or the Colorado General Assembly?'

# 5
# The Decline of the Party Structures

During the 1960s Democratic party organizations were under stress from several sources. There were problems of recruiting activists, of losing activists to candidate and single-issue campaigns, of rival campaign resources available through the growing profession of campaign consultancy, and of declining white populations in centre-city areas. These were sources of stress to be found in many Democratic parties. However, in addition to these widespread sources, there were four others, each of which had an important impact on some parties but not on all of them. The first was the legacy of the internal party disputes of the 1950s. Parties which had experienced conflict then found that this exacerbated the difficulties posed by disputes over the national issues prominent in the 1960s. It diminished loyalty to the party at the very time that the option of 'exit' to other forms of political activism was becoming easier. The parties in New York City were, perhaps, the most obvious examples of this in the country.

The second source was institutional reform. The 1960s and 1970s was an era of political reform comparable to that of the period 1900–20, but our understanding of these reforms is incomplete if we fail to realize that they resulted from two rather different political forces. One might be called 'democratic modernization' – changes to arrangements which were acceptable in industrializing society, but which were now incompatible with widely held social values. Thus the inability of state legislatures to respond effectively to demands for governmental intervention helped to bring about their professionalization; the attack on electoral malapportionment led indirectly to the more widespread use of single-member districts in state legislatures. We consider one of the effects of the professionalization of the legislatures in Chapter 6, while in this chapter we examine the adoption of single-member districts in Denver. Institutional reforms also emanated from much narrower political pressures, ones which were peculiar to the politics of the 1960s and 1970s. The transformation of the presidential nominating procedures is the outstanding example of this. Of course, none of the institutional changes was brought about exclusively by one force or the other – they involved the interaction of both. This is the case with several changes in nominating and election procedures in states which previously had arrangements to maximize party organization input – for example, Connecticut's abandonment of the mandatory party-ticket voting lever and New York's introduction of primaries for state-wide elections. In this study we are confining our attention to reforms which affected local party organizations and it is worth emphasizing that, although they have generated far more attention than other institutional changes, the presidential nomination reforms have probably had least impact on local parties.[1] They only came into operation in the 1970s, after party collapse was well established and after the rewards for supporting a winning presidential candidate were already in decline. Local and state parties were more affected, as they always had been, by losses of power over local and state offices than by loss of control over presidential nominations. Once we recognize this, we can see that the part played by amateur Democrats in reforming political institutions to the disadvantage of parties has probably been quite small.

A third source of stress arose in those states which did not have legally defined political parties below the level of the county. The absence of precinct organizations meant that, when the number of activists declined, there was no mechanism for providing organizational continuity until conditions for recruitment became more favourable. Voluntary clubs simply ceased to exist when they no longer had sufficient members, whereas the continuing existence of precinct committees made it possible that once again party could be used as a vehicle for mobilization. In part, the failure of the CDC clubs to revive at all in the 1970s was due to their extra-legal status: the club movement would have had to be created anew, and not merely re-activated. Instead, new organizational structures would emerge to channel the new concern with, for example, neighbourhood level issues. In the East Bay the particular form which these structures took was also influenced by the fourth source of stress – issue extremism. In the decade 1963–73 polarization on national issues produced political movements which fell outside the political consensus. In most cities these movements did not generate organizations to contest elections; their main effect on the Democratic party was the propaganda use made of them by Republicans who sought to link them to the Democrats. In one or two places, though, their intervention in elections did disrupt the established pattern of competition and transform it. The Black Panthers in Oakland and the radicals in Berkeley provided this kind of disruption, and these cases show how vulnerable were parties in cities where the legally defined party structure was weak.

In this chapter we examine these four sources of stress; each of the areas in our study exhibited at least one of them in an acute form. In the first section we consider the continuing intra-party conflict in New York City and the growth of disloyalty to the party. In the second section we examine how the Democratic party in Denver lost its power base when multi-member legislative districts were abolished and how it failed to develop an alternative base. Finally, in the third section we see how a combination of weak party structures and issue extremism resulted in party organizations in the East Bay being displaced from municipal elections.

## 1. New York: Fratricide and the Decline of Organizational Loyalty

The amateur insurgency of the 1950s did not always produce intra-party conflict which rendered party organizations incapable of performing their electoral functions efficiently. In Denver, for example, an accommodation was reached between the amateurs who took control of the party and the more conservative elements. Normally, the liberals would exercise restraint in challenging conservative incumbents, and in respecting the claims of 'senior' Democrats for major office nomination, while open competition was acceptable when state legislature seats became vacant. This tacit agreement survived until the influx of new activists at the end of the 1960s. But in New York, where the patronage stakes were much greater and there was little tradition of accommodating a reform element within the party, intra-party conflict adversely affected Democratic organizations in every borough. Unwilling to admit the reformers into the party, and fearful that their patronage operations would be undermined, the regulars were usually successful in slowing down transfers of power to the reformers. The major gains made by the amateurs in Manhattan, so that they became a substantial minority on that party's County Executive Committee by the early 1960s, were not repeated either there or in other boroughs after about 1965. However, the regulars paid a price for rearguard actions which safeguarded their control of the county party organizations. By concentrating their resources on protecting court patronage and the composition of the county parties themselves, they had fewer resources to influence elections to other offices. Increasingly, the regular organizations lost control over nominations to congressional and state legislature seats, and in doing so they allowed new rivals (the public office-holders) to establish themselves. Furthermore, both the regulars in defending their organizational base and the reformers in attacking it weakened the party itself, because increasingly they became willing to support candidates of other parties against Democrats. The ethos of party loyalty declined and this development was intensified by the peculiar role played by other parties in New York's electoral system. To explain this, it is useful first to outline the nature of the conflicts in the four larger boroughs.

While the Manhattan party is usually cited as the best example of intra-party conflict inducing powerlessness in the organization, it is the Bronx party which provides the best example of a 'no surrender' policy by regulars. This produced a disparity between the regulars' control over the party's internal affairs and their lack of power in the nomination process. Until 1978 the Bronx party systematically excluded reformers from the formal county organization. Even in the mid-1970s all the district leaders were loyal to the county chairman, a monopoly achieved partly by an alliance with one of the two main Puerto Rican factions in the borough and partly by the use of rules, which were unique in New York City, that made it difficult for 'grassroots' reform strength to be translated into votes on the County Executive Committee.[2] The main problem with this strategy, which was all too apparent to Stanley Friedman, the new chairman selected in 1978, was that absolute control over the organization had not stopped the election of non-organization candidates to public office. By 1977 two of the three congressmen, two of the four state senators, and five of the borough's eleven assemblymen were outside the former chairman's influence.[3] Far from being a reformer, Friedman was to use his office to 'wheel and deal' in the manner of his predecessors, but one of his first actions in office was to arrange a truce between the organization and the dissidents, so that in 1978 the only contested Democratic primaries in the Bronx involved individual dissidents.[4]

In fact, the real threat to the Bronx organization was not any growth in power by the reform movement but demographic changes. By 1980 about two-thirds of the borough was black or Hispanic and, faced with this in the 1982 reapportionment, the pragmatic Friedman decided to support proposals to increase minority representation in the state legislature.[5] In place of a rigidly hierarchical party, Friedman was attempting to preserve the influence of regulars through a series of alliances with certain minority politicians, some of whom had earlier been excluded from the party's inner circles. But whatever the value of pragmatism in the 1980s, it might be argued that the 'no surrender' policy of former years at least had helped to slow down the loss of public offices by regulars. Indeed, it must be admitted that, while newspaper articles in New York in the 1970s frequently discussed the waning power of the regular organizations, their loss of public offices in the Bronx was relatively small after 1967. In that year it was estimated that the regulars had already lost control of the Borough Presidency, three of the four congressional seats, two of the four state senate seats, and four of the twelve state assembly seats. However, to focus on this apparent success of the 'no surrender' policy is to ignore the growth in influence of individual opponents such as Herman Badillo, and later Robert Garcia, which increased with their tenure in public office. In the long term, then, the position of the regulars was weakened by a failure to accommodate such leaders; this maximized the potential for coalition building against the regulars. In the short term, though, the policy could be judged a success, in that most of the losses of public office were confined to the period from 1963 to 1967. What lay behind this short-term success for the regulars were divisions in the minority communities and a failure by white reformers to expand beyond their original power base. As we saw in the last chapter, amateur and regular clubs in New York experienced declining recruitment during the late 1960s and the 1970s, and in the Bronx this setback for the amateurs was coupled with the great difficulties they encountered in developing alliances with factions in the minority communities.

Brooklyn, sometimes heralded in the early 1970s as the most powerful Democratic county organization in the nation after Chicago, did not experience the more ruthless use of organizational power found in the Bronx.[6] Under Meade Esposito traditional local party activities were continued and, while electoral contests with non-regular Democrats were not eschewed, the party did attempt to limit public displays of conflict. However, as in the Bronx, the 'minority issue' was dealt with by alliances with those leaders who could most easily accommodate themselves to the regulars' ethos. This prevented the growth of a serious threat to the organization itself from elements, but it did not stop the party from losing its ability to control nominations to Congress. With one exception, the Brooklyn congressional delegation in 1970 was highly supportive of the party organization; the exception was Shirley Chisholm who, if she was not quite as independent as she claimed to be, was not an ally of the regulars. By 1974 the situation was very different.

Of the six members of Congress, two (Holtzman and Solarz) were associated with the reform movement, one was Chisholm, and another (Scheuer) was a wealthy ex-congressman who had little need of the party's electoral resources. Even the two remaining members were scarcely old-style party types – one was an ex-policeman and the other was a wealthy businessman with both reform and regular connections. In brief, this was not the kind of delegation that could have been treated in the notorious manner congressman Dominic Daniels was supposed to have been treated by the boss of Jersey City in 1961. When it appeared that Daniels might defect on a vital vote on aid to Catholic schools, the White House contacted Mayor Kenny of Jersey City who is supposed to have said to his congressman: 'Who sent you there, me or the Bishop? And who's going to keep you there, me or the Bishop?'[7]

If the Brooklyn congressional delegation in the mid-1970s was not the puppet of Meade Esposito, however, he was more able than the Bronx leaders to preserve party control over nominations to lesser public offices. In addition to his power over judgeships in the borough, at the beginning of the 1970s Esposito controlled the Borough Presidency, the District Attorney's office, and most of the state Assembly, state Senate, and city council seats. During the next decade there was a decline, but not a spectacular one, in his nominating power. The two most significant losses were in this alliances with black leaders which cost him a few state legislature seats, and, in 1981, in the election for District Attorney. As in the Bronx, the problem for a white machine leader was that a growing non-white population was making demands for services which lay outside the domain of the machine, and which could be satisfied only partially by the black leaders whom the bosses had made their allies. At best, then, these alliances slowed down their loss of power to those outside the machine. Paradoxically, though, the loss of the District Attorney's office emphasized the advantages the machine still enjoyed over its activist opponents at the beginning of the 1980s. This election was won by former congresswoman and US Senate candidate Elizabeth Holtzman, who was able to build a coalition between white reformers and non-machine blacks precisely because she was a well-known candidate. Normally this coalition had proved elusive to those who had attempted to construct it. In 1981 the machine lost the nomination because someone of Holtzman's standing decided that she wanted it; she put together the coalition, rather than inherited it.

Possibly the best exemplar in America of the potential for survival by a regular organization in the face of extensive amateur mobilization was the Manhattan party. The reformers were already an important minority in the borough in 1962, when *The Amateur Democrat* was published, but it was not for another fifteen years that a reformer became chairman of the party. (Even then, the accession of Miriam Bockman was that of an older style of politics; she was the beneficiary of the convention that a new mayor can depose the party leader in his home borough. In this instance Edward Koch secured the resignation of Frank Rossetti, and lent his influence to the selection of Bockman, the leader of Koch's old club (the Village Independent Democrats), as the replacement.) There were three main causes of the amateurs' failure to take over formal control of the party organization earlier than they did. They failed to satisfy the demands of the Harlem leaders in the years after Adam Clayton Powell's decline; there was internal dissent among the reformers, not only in the West Side around 1970, but also more generally between East and West Siders; and when necessary, the regulars were able to mobilize resources for local election contests. This last point is a significant one; even though the reformers eroded much of the patronage base of the county party, the regulars did not concede to their opponents. Epitomizing this last problem for the reformers, though in no respect was it a typical election, was the 1971 contest for district leader in East Harlem between party chairman Rossetti and the millionaire publisher of *Village Voice*, Carter Burden. It was not typical because even by 1971 fiercely contested fights between regulars and reformers for district leaderships were becoming rare, and because few reform candidates had the financial reserves of Burden who was alleged to have spent $15,000. It might have seemed an unpromising contest for Rossetti, who would have lost his county chairmanship automatically if Burden had become district leader, because his Italian electoral base had been eroded and the district was rapidly becoming a Puerto Rican dominated one. However, the stakes were seen by several leading Democrats to be sufficiently high to warrant an unusually well-financed campaign; Rossetti was reputed to have spent at least $40,000 in the campaign, and thus been able to employ campaign workers from other districts.[8] Rossetti survived this challenge and remained borough leader until 1977. Nevertheless, his position was very different from that of the Bronx or Brooklyn leaders. He was tolerated by at least some reform leaders because they did not believe that there was anyone else who was more capable of co-ordinating the fragmented elements of the party. Moreover, unlike the Bronx and Brooklyn leaders, the patronage power of the Manhattan chairman had been reduced both by the size of the reform group on the County Executive Committee and by rules relating to the selection of judges. But the most important difference between their positions and his was that continuous internal strife made it very difficult to use the party organization as a vehicle for *any* co-ordinated activity. From being the leading borough party after the city's consolidation, Manhattan's influence was in decline in the 1950s, and by the 1970s it was arguably the weakest borough party. It was not weak because of the large amateur presence, but because internal conflict continued long after the substantive disputes – over public policy and party procedures – had declined.

The parties in Queens, and in the small (and thus politically insignificant) borough of Richmond had long been controlled by regulars. Their much larger white populations prevented the organizations from having the kinds of alliance problems with minorities which were afflicting the parties in the other three boroughs in the 1970s. In Queens reformers made a few gains from the regulars in much the same way as their counterparts in Brooklyn did: they could control a few districts, but they never came close to controlling the county organization. In those districts they controlled the reformers could act independently of the county party. One of the best examples of how organizational control was becoming inadequate for controlling election outcomes was an incident in 1974. That year Queens party chairman Matthew Troy and the party's Executive Committee announced that they wanted to restore strict party discipline and that they intended to redesignate all incumbents except one; gubernatorial candidates were asked not to endorse anyone other than 'official' candidates. The one incumbent who was not redesignated was a city councilman who had voted for a homosexual civil rights bill which Troy had opposed. There were two interesting aspects of this attempt to restore party discipline. The first was that Troy did not attempt to oppose another councilman who had also voted for the bill, because he was a reformer with a secure electoral base in his own district. To have attempted a display of power there would have been counter-productive. The second aspect is that the attempt failed; the threatened incumbent won both the party primary and the general election that year.[9]

What had begun as a revolt by liberals in Manhattan in the 1950s had become a city-wide phenomenon by the second half of the 1960s. Yet, as we have seen, even in Manhattan the successes of the early 1960s did not lead to a seizure of power by a cohesive group of liberal activists; rather, the party atrophied in that borough with internal conflict. As we saw in Chapter 4, there were several reasons why the growth in the reform movement should have been curtailed after about 1965–6. Some were national political developments, such as the splitting of the amateur movement over the Vietnam war and the 'exit' by amateurs from clubs and party organizations to individual campaigns. There was also one local development which played a major role in reducing activist recruitment to the party and in reducing loyalty to the party. This was the election in 1965 of liberal mayor John Lindsay on the Republican and Liberal party tickets. After the Democratic primary many of the amateurs preferred Lindsay to the Democratic candidate Beame, but their late and often muted expressions of support for him meant that they were not a central part of his coalition. Among political liberals the Liberal party members received the main spoils of patronage. Thus, if regular Democrats were denied patronage during the Lindsay years, at least until he changed his party registration in 1971, neither did the reform Democrats have 'the ear of the mayor'. This left the amateur movement in an ambiguous position: unlike the regulars, they generally supported the administration, but had little direct influence on it. Denied a coherent position with respect to whether they were 'government' or 'opposition', amateurs were prone to division among themselves.

The complexity of the continuing intra-party divisions encouraged candidates to rely more heavily on their own organizations to perform electoral tasks, rather than on bodies which might have objectives other than maximizing party success. But there was another, equally important, consequence of the factional complexity the Lindsay years produced. Increasingly, regulars and reformers became disloyal to the party when it was convenient to do so, and would support, and sometimes campaign for, non-Democratic candidates at a general election. Party loyalty had never been absolute, but in the past disloyalty had been connected with major, long-term shifts in the party power structure. The disloyalty of the 1970s was both more frequent and more pragmatic in character; it was much less connected with struggles inside the party organizations. In part, disloyalty was encouraged by a process unique to New York – cross-endorsement of major party candidates by third and fourth parties, the Liberals and Conservatives – but cross-endorsement was by no means wholly responsible for this. The extent to which disloyalty became more acceptable can be gauged by comparing Edward Koch's experience in 1965 with that of Herman Farrell in 1981. In the 1965 mayoral election Koch and his fellow district leader gave a late endorsement to Lindsay. Several members of the county party's Executive Committee demanded their expulsion; even some reform leaders were in favour of their censure. In the event, a rather ambiguous statement about the need for party loyalty was agreed upon, and the two rebels kept their seats on the committee. In 1981 the Manhattan party chairman, Herman Farrell, himself endorsed the Liberal party candidate for Borough President over the Democratic nominee after the latter had defeated the former in the Democratic primary. Farrell was never in any danger of being even reprimanded for his action, despite the fact that the Liberal candidate subsequently lost the general election.

Disloyalty has taken many forms. One of the most common has been to make use of the cross-endorsement procedure, either to get on to the general election ballot after defeat in a major party's primary, or as a means of eliminating major party opposition at the general election. The best known instance involving the Democratic party in the first of these strategies was that of Mario Cuomo in the 1977 mayoral election. Cuomo, who the following year was to become the Democratic Lieutenant-Governor of the state and its Governor in 1982, entered the Democratic primary but also received the Liberal party nomination. After losing the run-off election to Koch, Cuomo maintained a serious challenge to him as a Liberal at the general election and again lost. Minor party ballot lines are not always unsuccessful. John Lindsay, of course, retained his office after losing the Republican mayoral primary in 1969. A less well-known instance is that of Elliot Engel, a reform district leader in the Bronx; in a special Assembly election in 1977 Engel ran as a Liberal to defeat the nominee of the regular organization. Generally, however, minor party lines on their own are not likely to bring victory; their real value is that capturing them restricts the number of potential rivals whose votes might help a Republican opponent. The second strategy, that of capturing the Republican nomination, is a more radical step in the disarmament of opposition. Among the more prominent Democrats who utilized this were congressman Biaggi and, in 1981, Mayor Koch. It is a form of disloyalty which directly benefits the individual candidate, without immediately harming his party, though one of its long-term effects is to reduce the salience of party for the electorate.

If capturing the opposition was one form of disloyalty to the party, another form which developed was the sustenance of that opposition. Both individual public office-holders and regular Democratic leaders have used this device as a way of minimizing the combined opposition forces to them, both inside and outside the party. On several occasions in the 1970s the chairman of the Bronx party endorsed a Republican state senator in his borough, and he even gave that endorsement in 1974 when the Democratic party state-wide hoped to take advantage of the Watergate scandal and win control of the state Senate.[10] By 1980 such behaviour was widespread. That year the party chairmen in the Bronx, Queens, and Staten Island arranged for the endorsement of three Republican state senators, while the Queens chairman also allowed another Republican senator to enter the Democratic primary.[11] Democratic mayor Koch endorsed two of these Republicans and, in doing so, incurred the public displeasure of both the Democratic leader in the Senate and the Governor. Ironically, the former was not an exemplar of party loyalty himself, for he had publicly opposed the party's mayoral nominee in 1969.

It should be noted, however, that there is yet a fourth way in which New York's third and fourth parties have proved hostile to party politics, though it is the Republicans who have been most affected by it in recent years. An early endorsement of a candidate by one of the two lesser parties could place him in the best position to capture the nomination of a major party; in effect, a minor party could be the most important influence on the decision of a major party. The decline of the Liberal party has given it no such opportunities in relation to the Democrats, but the Conservatives have been decisive in at least two Republican contests. In 1980 they endorsed a little-known county official, against incumbent Republican US Senator Jacob Javits, and Javits lost the Republican primary. In 1982 the Conservatives' early endorsement of an equally unknown businessman enabled him to secure the Republican gubernatorial nomination.

One of the main themes of this book is that party decline in the 1960s and 1970s was brought about by a number of different causes. The growth of party disloyalty in New York is but one example of the complexity of the processes involved. Disloyalty was not simply the result of the internal conflicts of the 1950s and 1960s, although they contributed to its development; it was also facilitated by institutional changes outside the control of the New York Democratic parties. The most important change concerned nominations to major offices. After 1967 nominations to the US Senate and the state governorship were no longer decided by agreements among bosses in conventions, but in primary elections. Under pressure from 'good government' lobbies, the legislature and Governor Rockefeller had produced a reform of state law which was supposed to retain a major role for the parties in designating candidates, while allowing challengers with significant support in the party to force a primary election. In fact, most nominees came to 'be selected not by the state committees but rather in free-for-all primary contests'.[12] At a stroke, ill-conceived legislation changed New York from one of the last bastions of the convention system to a state in which the individual candidate for state-wide office became the central actor in the nomination process.

In the late 1960s and early 1970s the absence of Democratic leadership in the state was an additional factor which exaggerated the impact of intra-party conflict in New York City. At that time both the state Governor and the city mayor were Republicans and, after Robert Kennedy's assassination in 1968, the two US Senators were also Republicans. But to acknowledge the institutional and leadership factors in accounting for the party's decline is not to admit that the changes we have described would not otherwise have occurred, rather it is to explain the particular framework within which the party's collapse took place. For example, the stabilizing influence Kennedy brought as the party's *de facto* leader after 1965 was that of an 'outside' individual candidate who had largely imposed his nomination on the party; it is best conceived as an aspect of the transformation the party was undergoing, and not as an element of the old party style. Yet, with respect to campaigning, the difference between the mid-1960s and the late 1970s is as striking as the difference between the former period and the pre-Wagner era. It was in Wagner's time that the relationship between the candidate, especially an incumbent, and the party leaders began to shift in favour of greater autonomy for the candidate. It was Wagner, in 1961, who was the first Democratic mayor to run 'against the bosses' at an election. However, a high degree of 'team work' and a 'team ethos' still remained within the party. Candidates on the same ticket endorsed each other, and party leaders who opposed members of the ticket were not supposed to voice opposition after the primaries, let alone do what Koch did in 1965. Moreover, opposition within the party was factional opposition; in 1965, for example, Paul Screvane had his own ticket in the primary election when running unsuccessfully for the Democratic mayoral nomination. Candidates for the City Council Presidency and for City Comptroller ran as a team with Screvane.[13]

By the late 1970s candidates campaigned mainly as individuals, even though groups like the NDC would endorse a slate of candidates in the primaries and the clubs would distribute their own literature. Furthermore, as we have seen, candidate and organizational loyalty to fellow bearers of the party label had declined to the point at which endorsements might be withheld or given to candidates of other parties. This did not mean, though, that a system of team work had been replaced by one of non-interference in the campaigns of others; there were still many circumstances in which a candidate would see interference as advantageous to him. One instance of this occurred in the Koch campaign for the governorship in 1982. As the leading candidate for the Democratic nomination, Koch believed he was in a position to endorse the stronger candidate for the Lieutenant Governor nomination. Even so, his public involvement in another nomination contest stopped well short of creating a 'slate', let alone a 'tight slate' of the kind found twenty years earlier. While there were advantages to Koch in dissociating himself from the weaker candidate, there were no such benefits to the candidate he endorsed in becoming too closely identified with Koch; indeed, Koch unexpectedly lost his primary election, while the person he endorsed won. If electoral politics in New York has not become completely atomized, with candidates concerned only with their own contests, equally it is no longer characterized by well-organized factions within parties and battles between tightly-knit teams of parties.

We have seen that in New York the parties continued to play some part in the nomination and election process, but that they had declined in importance *vis-à-vis* the individual candidate. By the end of the 1970s the relationship between candidates and clubs was approaching that found in California at the height of CDC influence in the 1950s. But the rise of the individual candidate did not mean that candidates refrained from interfering in the elections of others. This was one of the major differences between New York at the end of the 1970s and California in the 1950s. The history of organized interference in New York meant that the practice persisted even as the candidates were becoming the dominant actors in the electoral process. As we shall see, this was very different from the transformation in candidate–party relations which was taking place in Denver.

## 2. Denver: Institutional Reform and Non-Reform

In Chapter 3 we saw how a relatively strong political party organization was established in Denver through the central role that the party played in the nomination and election of the city's state legislators. This role had been made possible largely by the use of a city-wide multi-member constituency for both chambers of the state legislature. In the late 1950s Denver was not unusual for an American city in employing a multi-member constituency for the lower chamber, but it was unusual in using them for Senate elections and in having so many assemblymen elected from one district. In 1955 the use of multi-member districts in America had been analysed in an important, but often overlooked, study by Maurice Klain.[14] Klain wished to expose a popular myth – that the single-member constituency was the method of electing legislators typically adopted by the American states. It was not. Only nine of the forty-eight states used this method for every legislative seat. Thirty-six states had some multi-member districts in their lower chambers, and of these states only in Iowa did the multi-member seats constitute less than 25 per cent of the total number of seats. In twenty-one states at least half of the seats in the state House were contested in multi-member districts. Klain argued that it was the single-member district which was the more recent practice, and that in earlier years there were fewer states using only this method. However, single-member districts were much more common in state Senate elections; in 1955 88 per cent of all Senate seats were contested in this way, although thirty-two states made some use of multi-member districts in these elections. Since then there has been a considerable change in the way candidates are elected to lower chambers, but relatively little *aggregate* change in elections to upper chambers.

In the case of state Houses, the magnitude of the change is concealed by the relatively small increase in the number of seats contested by the single-member district system. In the early 1950s 55 per cent of all seats were contested by this method, while by the late 1970s 64 per cent of them were. But, whereas only twelve states used nothing but the single-member system in the 1950s, twenty-eight of them did so two and a half decades later. Eighteen states abandoned the use of multi-member districts entirely while only two states started to use them, although both Alaska and Hawaii adopted this method on their entry to the federal union and continued to use it. Correspondingly, there was a great increase in the number of states using nothing but single-member districts for both chambers – from nine to twenty-six. In the case of state Senates, the number of seats contested in multi-member districts actually increased as a proportion of the total (from 12 per cent to 14 per cent), although the number of states using such districts declined from sixteen to thirteen. While ten states were abandoning their use for these elections, five states (in addition to Alaska and Hawaii) were adopting them. Nevertheless, the general trend among state legislatures was markedly towards the uniform use of the single-member system. The popular myth which Klain sought to expose was becoming a reality twenty five years later.

A distinct regional pattern has emerged in the use of the single-member district. While about half the states employ only this system, these are states which between them contain nearly two-thirds of the nation's population. With some notable exceptions (six, to be precise), the multi-member district has been retained primarily by southern states and by small states in the extreme north-east and west of the country. As can be seen in Table 5.1, only a quarter of the larger states outside the south still made use of the multi-member system in the 1970s. This proportion remains the same if we restrict our attention to the eight largest non-southern states: only two (Illinois and New Jersey) employed this system. However, among those states which retained it, the system was used extensively; eighteen of the twenty-two states employing it in lower chamber elections elected at least half of their members with it. This helps to explain why the aggregate increase in single-member districts in the country is relatively small, at 9 per cent of the total number of seats.

The regional pattern which is now to be found in the distribution of multi-member districts among the states did not exist in 1955. Admittedly, the nine states which used only single-member districts were not a microcosm of the nation – there were no southern states, and the group included both California and New York. But the rest – Delaware, Kansas, Kentucky, Missouri, Nebraska, Rhode Island, and Wisconsin – differed greatly in size, and were unusual only in that they included no other western state.

The adoption of the single-member district by most of the medium-sized and large states outside the south was not accidental. Although there were many local factors which help to account for the abandonment of, or continued use of, multi-member districts in particular cases, there were two national developments which promoted the adoption of the single-member district. One was the increase in party competition in the states. This began with the New Deal realignment and the breaking down of the 'system of 1896'.[15] Obviously, this process is taking decades to work through the political system, but in many non-southern states full-scale two-party competition was evident by the late 1950s

**Table 5.1**: Use of multi-member districts for State Legislature Elections by number of Congressional Districts – 1978

| | Non-southern states | | Southern states | |
|---|---|---|---|---|
| | Only single member districts | Some multi-member districts | Only single member districts | Some multi-member districts |
| States with 5 or more congressional districts | California<br>Colorado<br>Connecticut<br>Iowa<br>Kansas<br>Kentucky<br>Massachusetts<br>Michigan<br>Minnesota<br>Missouri<br>New York<br>Ohio<br>Oklahoma<br>Pennsylvania<br>Wisconsin<br>(15) | Illinois<br>Indiana<br>Maryland<br>New Jersey<br>Washington<br>(5) | Alabama<br>Louisiana<br>Tennessee<br>Texas<br>(4) | Florida<br>Georgia<br>Mississippi<br>North Carolina<br>South Carolina<br>Virginia<br>(6) |
| States with fewer than 5 congressional districts | Delaware<br>Montana<br>Nebraska<br>New Mexico<br>Oregon<br>Rhode Island<br>Utah<br>(7) | Alaska<br>Arizona<br>Hawaii<br>Idaho<br>Maine<br>Nevada<br>New Hampshire<br>North Dakota<br>South Dakota<br>Vermont<br>West Virginia<br>Wyoming<br>(12) | | Arkansas<br>(1) |

and early 1960s. Multi-member districts were a device which aided the dominant party in a state; they diluted the voting strength of the other party, through forcing the latter's candidates to compete in large districts against the numerically greater electorate of the former. Once they could expect to control state government on a regular basis, the leaders of the minority party considered dismantling an advantage enjoyed by their opponents. Of the thirteen non-southern states which converted to single-member districts after 1955, ten were rated by Austin Ranney as being two-party competitive in the period 1948-63; of the seventeen which retained multi-member districts, only nine were rated as competitive.[16]

The second development was the entry in 1962 of the US Supreme Court into an area it had previously deemed to be a 'political' matter – the apportionment of electoral districts. Earlier Courts had steered away from such politically sensitive issues, but under Earl Warren this policy was reversed.[17] Utilizing the principle of 'one man one vote', the Court ruled unconstitutional voting arrangements which did not provide the same power to each vote. For the first time states had to conform to a national standard, and in many cases the preservation of multi-member districts would have made conforming too complex. These states tended to opt for the administratively easier task of drawing boundaries for an entirely single-member system, especially when there were large concentrations of non-white minorities. Since the multi-member district was known to be a device which could be exploited in reducing the voting power of minorities, states which did not wish to circumvent the Supreme Court principle, or did not want to risk legal challenges to their apportionment schemes, abandoned multi-member districts. Correspondingly, those states which did not have large urban minority populations had less incentive for doing so: only four of the fourteen smaller non-southern states which employed multi-member districts in the 1950s had abandoned them by the late 1970s.

The advantage to a dominant political party in having to contest elections in multi-member districts extends beyond the diminution of its opponent's voting strength. In two ways this system provides for greater party control over its own candidates. First, it reduces the efficacy of individual campaigning in a primary election, and thereby makes the candidate more dependent on the party and its organization members in securing his nomination. Secondly, it reduces the role of individual campaigning at the general election, and correspondingly increases that of party or slate campaigning. It is not surprising, therefore, that three of the five large non-southern states which retained multi-member districts are states in which 'bossism' of some kind has survived. In addition to Illinois, there is New Jersey with its partially collapsed urban machines and Indiana with its state-wide system of financial contributions by government employees to their parties. It must be emphasized, however, that districting arrangements alone do not produce strong parties. Other electoral institutions, laws, or practices can negate, or greatly reduce, the effects of a multi-member system. Among the users of this system, we find not only states such as Indiana but others such as Washington – a state which has had weak party organizations. Moreover, the potential for party control varies with the size of the electoral district: it will be much less in a two-member district than in a seventeen-member district. For this reason it would be absurd to claim that the Denver experience was typical. It was not; the city had one of the largest legislative districts in the country. Thus, the Denver case will show in an exaggerated form the effects of the move away from multi-member districts. This case is interesting because it shows how party organization collapse could occur even in a party which had not had problems in recruiting activists or in containing intra-party conflicts. It is also an example of institutional reform for which amateur activists cannot be held responsible; whatever the merits of the argument that amateur politics destroyed the presidential party through reforming it, in Denver the amateurs were not the cause of party breakdown.

Single-member districts were first introduced in Denver in the 1964 election, as a result of the Republicans gaining control of the state in 1962. Unlike the more usual instances, the move away from multi-member districts in Colorado was not the consequence of newly-emerging party competition, but was an attempt to upset a long-standing balance of voting strength between the parties in one of the most competitive states in the country. Because of a court case, multi-member districts were restored for the 1966 election but were re-established permanently in 1968.

The effect of this measure was to change the optimal strategy for a state legislature candidate, both in trying to win his party's nomination and in campaigning at the general election. He was dependent no longer on the county party organization in determining his ballot position in the primary election; designation was now entrusted to those delegates from his own district. While in theory this should have increased the power of individual district captains, in practice it did so in only a few instances. The reason for this was the small size of the districts; in the case of House seats, each of them had a population of less than 35,000. By simply mobilizing friends and neighbours, any candidate would have a reasonable chance of getting the 20 per cent of the vote he needed at the assembly to gain access to the ballot. Moreover, the small size of the districts made a candidate's ranking on the primary ballot less important than it had been; it had now become relatively easy for a candidate to appeal 'above the heads' of opponents in the party to the electorate. Thus, the direct result of the adoption of single-member districts was not decentralization of power to the party at the district level, but an increase in the power of legislative candidates *vis-à-vis* the party at all levels. Of course, district captains were not always, or even usually, bypassed by candidates when mounting a campaign for the party's nomination, for the captain was still a useful ally. But the captains could not control the nomination process, even in the old 'ethnic' neighbourhoods of north-west Denver. With this loss the county party also lost its ability to raise funds from the candidates for the party campaign. With single-member districts, county-wide campaigning was only of concern to the party organization; it was of purely secondary interest to the individual legislative candidate. There was now little incentive for candidates to contribute a large portion of their campaign budget to the county party. They did not owe their nomination to that body and, whereas the 'return' on any dollar spent by the county party was difficult to estimate, it was all too easy to see how the marginal dollar could be used to improve their own campaigns. In this way the county party lost one of its principal sources of campaign funding and its role as the major source of Democratic campaign literature in the county. It also acquired more competition for the use of campaign activists. With the multi-member system the relatively small campaigns mounted by individual state legislature candidates left the party with only candidates for major state-wide offices and congressional seats, as rivals for campaign labour. Now it faced direct competition from seventeen other candidates.

If in retrospect we can see that the key element in providing for party organization vitality had been removed, we must not infer that this was apparent immediately to all participants. It was not, and there were several reasons for this. The first is that the party's openness and its methods for selecting delegates to the National Convention made it an important vehicle for opponents of the Vietnam war. 1968 was a turning point for attendances at party caucuses: the numbers participating in them increased greatly that year, and remained at a high level for the next few election years. It was difficult to believe that the party was disintegrating when participation in some of the nomination processes was greater than ever. Furthermore, at least until 1974, candidates at all levels of office continued to take seriously victory in the nominating assemblies, rather than relying primarily on an appeal to the primary electorate. The turning point that year was Gary Hart's US Senate campaign which demonstrated the increased value of mobilizing activists from outside the ranks of the party.[13] This too did not discredit completely the role of the party faithful in nominations for other offices, for the office Hart sought was the one for which 'outside' mobilization was arguably easiest. Nevertheless, even if by 1974 candidates were still placing some value on an assembly victory over their opponents, they had already come to rely much less on the party for campaign services. After 1966 partisan campaigns increasingly became candidate-centred. However, there was another factor which disguised party weaknesses in the late 1960s – the party's close involvement in local elections.

Despite the failure to replace the non-partisan system of elections in 1960, both major parties remained important participants in the city's mayoral elections. Once again in 1963, as in 1959, the absence of a party primary led to two Democrats competing in the first election, but the party united behind the surviving candidate for the run-off election, and then campaigned as if the office was partisan. From 1963 the mayor, Tom Currigan, was a party-oriented Democrat, so that the party at least had an important link with a new source of power just as its original source was being removed. But for an unusual series of events that link would have become a more formal one in the late 1960s.

In 1968 a referendum amending the city charter to make mayoral and city council elections partisan contests was passed, though in circumstances which were particularly favourable to its proponents. Shortly afterwards Mayor Currigan resigned to take a business post in California. Negotiations to prevent the conservative Democrat, William McNichols, his constitutional successor, from assuming power failed and McNichols duly became mayor. Initially the main disadvantage of the arrangement to the Democratic party appeared to be that McNichols would serve in the office for two and a half years before he could be replaced. On technical grounds, however, a court in 1969 declared the referendum on partisan elections void. Undoubtedly, McNichols was the principal beneficiary of the court's decision, for in a

Democratic primary he would have struggled to defeat a more liberal challenger. With Republican voters not excluded from the first round of the mayoral election, he could seek to build a bipartisan moderate-conservative coalition – a task he undertook successfully in 1971, 1975, and 1979. An era of quasi-partisan mayoral elections had ended. But what was the Democratic party's response to the McNichols 'coup'?

There can be little doubt that by 1969 Democratic party activists expected local election campaigns to be highly partisan in style. Ever since 1955, when reformer Quigg Newton left office, the mayoral campaigns had involved large-scale efforts by both the Democratic and Republican organizations. Currigan's years as mayor had brought the Democratic party into local government, and the partisan element in local elections was strengthened in 1969 by the party's first overt participation in Denver school board elections. It was scarcely surprising, therefore, that the party's response to McNichols in 1971 was the most explicitly partisan campaign ever. The former county party chairman, Dale Tooley, ran against McNichols and the new party chairman, Richard Young, obtained for him the formal endorsement of the County Executive Committee. From one viewpoint the party's involvement in the campaign was little different from earlier years. But from another perspective Young's approach represented a radical departure. The party organization was being used against another Democrat, whereas in the past the organization had not become involved until there was only one Democrat left in the field. Moreover, the Executive Committee's formal endorsement of Tooley was an isolated instance of proceeding as if the city's elections were really partisan; no attempt was made to endorse candidates for other offices that year. Indeed, after the defeat of its endorsed candidates in the 1969 school board elections, the party abandoned its policy of overt participation in those elections because it was deemed to be counter-productive. As a result, the Tooley endorsement was widely interpreted as being a device that one faction of the party was using against another. The significance of this was not that it discredited the party with the electorate – where the impact was probably small – but that it helped to reduce the commitment of McNichols supporters in the party to that organization. A split between liberals and conservatives, created mainly by the successful challenge to congressman Byron Rogers in 1970, was made worse by the leadership's action against McNichols. It undermined party loyalty, and thereby encouraged the growth of loyalty to individual candidates. Of course, we have no way of measuring the decline in activist loyalty attributable to this manoeuvre, but the failure to elect Tooley in 1971 was a watershed for the party. Before 1971 the party had been at the centre of campaigning for many partisan and non-partisan offices, though, as we have seen, its role was being threatened. After 1971 the party played, at most, a secondary role in every campaign – with the exception of a repeat election between Tooley and McNichols in 1975, where party involvement was largely determined by what had happened four years before.

In Denver, then, the failure to find a new base of institutional power after the abolition of multi-member districts worked in combination with recent intra-party conflict to provide an incentive for activists to 'exit' from party work. If, unlike New York, there was not a long history of conflict which might induce disloyalty, the impact of the one dispute was, perhaps, more devastating than it would have been in a city such as New York, because the party now had so few incentives with which to induce loyal, party-oriented, activism.

One irony of the party's demise after 1971 was that Richard Young wanted to expand the organization's role in nominating and campaigning for candidates. In 1975 he was responsible for an attempt to involve the party in the elections for the two at-large city council seats. These seats had been created in 1971 and, unlike other municipal offices, no provision had been made for a run-off election. Consequently, if there were a large number of candidates, the winners could be elected with a small share of the vote. In 1971 there were more liberal than conservative candidates, and Young and other Democrats saw this as the cause of a failure to elect a liberal Democrat. Determined not to have a repetition of this in 1975, Young proposed that the party's Executive Committee be used to reduce the number of potential candidates. The Committee would act as a nominating convention by endorsing two candidates for the at-large seats.

Young's objectives were largely attained: there were fewer candidates than in 1971 and one Democrat was elected. Nevertheless, the party's role did not extend beyond the nomination process. The two endorsed candidates organized their own campaigns after the unofficial convention, and the party did not raise or spend money on their behalf. Even this limited expansion of the party's role had been made possible because there were no incumbents whose views had to be considered. For any incumbent seeking re-election the party endorsement procedure had virtually no advantages. Given the low visibility of these elections, a profusion of candidates increased the election prospects of any candidate who was well-known – and usually incumbents would be the best known. Thus, few incumbents could ever be expected to withdraw from a campaign merely because they failed to obtain an endorsement, even if it was that of a political party. This flaw in the informal arrangements became evident in 1979 when the incumbent Democrat, who had become more conservative in office, refused to submit herself for endorsement and no convention could be held. The result was that party influence on the election was negligible. Young's efforts four years earlier could not establish a precedent for party involvement, because the party had nothing to offer the one actor whose co-operation was essential. In the absence of a legally-defined role in the nomination process, the party could not intervene permanently in that process – at least, it could not if it was not a major contributor of campaign resources to the candidates.

The two most important resources the party lacked in non-partisan elections were money and control over the organization's personnel. There was no regular source of income available to the party which it could use to be a major contributor in elections. Through its power over the nomination of state legislators before 1964, the party had created a biennial flow of funds from these candidates which was sufficient to generate a city-wide campaign when they were added to those provided by its other three sources. There was no surplus, however, which could be used in other elections. Moreover, unlike Chicago, Nassau County (New York), and Indiana, there had never been any provision for public employees to pay part of their monthly salaries into the party coffers. Nor does there seem to have been much *direct* income from the party's judicial connection. In brief, with the assessment system ceasing in 1966 and Rogers raising money for the party for the last time in 1968, the Denver party organization of the 1970s was now dependent on voluntary contributions. As this kind of fund-raiser, it had to compete with the individual candidates; in elections of all kinds the party was at a disadvantage, as potential contributors often believed that they could be more sure of what they were 'buying' in the case of a candidate rather than of a party.

The party's problems in mobilizing the precinct committee people for city elections were almost as great as those it experienced as a financial contributor, and indirectly they derived from the party's lack of income. Relying on their own efforts in raising campaign funds, candidates for mayor, city auditor, council, and school board did not campaign jointly or as part of a slate – there was no incentive for co-operation. In the absence of even informal coalitions, the party could not use the sort of persuasion and peer group pressure on party activists which might have been partially effective with those who refused to work for the Democratic ticket in a partisan election. Of course, party leaders could try to encourage organization members to work for a particular candidate, but the most effective agent for this kind of mobilization was the candidate himself. Campaigning by party slate might have increased party influence in the mobilization of campaign activists, but there were two reasons why it did not develop. First, there was the legacy of the Stapleton era. Part of that legacy was public acceptance of a quasi-partisan mayoralty, which permitted parties to support individual candidates. In run-off elections straight competition between Democratic and Republican backed candidates was acceptable, and this was the usual kind of mayoral election in Denver in the 1950s and 1960s. But, correspondingly, there was no tradition of parties overtly seeking to restrict the field of mayoral candidates before the run-off stage. The absence of this tradition was a major reason why partisan elections were considered by Democratic party leaders to be so important in expanding the party's role. Secondly, to have bucked tradition, and formally endorsed an entire slate early in the campaign, might have appealed to some party leaders if all the offices had the same electoral constituency – but they did not. While all other officials represented a city-wide constituency, the nine council members elected in districts (eleven after 1971) represented different areas of the city. Any attempt to have given a party endorsement in these elections would have raised the problematic issue of who should be responsible for such an endorsement – the county party or the party organization in each district. Given the ideological and ethnic diversity in the city, it is unlikely that a slate could have been constructed which satisfied both particular and majority interests. In the face of potential conflicts with local organization members, the county party never developed a close relationship with candidates for the district seats, and it was left to the candidate to initiate any involvement in his campaign by members of the district party organization. The introduction of the two at-large council seats in 1971 did not change the party's non-participation in the recruitment of candidates and in campaigning for the district seats.

A decade of unsuccessful attempts to define an acceptable role for the party in non-partisan contests reached its nadir in the 1979 city elections. The party intervened in only two elections. In the contest for city auditor the county party made an endorsement in the run-off election – the recipient being the conservative incumbent who subsequently lost to the Republican-endorsed candidate. But it was in the contest for the two posts on the city's Election Commission that the party revealed the extent of its decline. An incumbent Democrat was running for re-election and he received the party's endorsement, but the party made no attempt to recruit a second candidate. One Democrat was now running against two Republican-endorsed candidates, and many Democratic voters did not understand the strategic advantage of not voting for a second candidate and they voted for one of the two Republicans. As a result both Republicans out-polled the Democrat, when even a token candidacy by another Democrat might have kept him in office.

In many ways the collapse of Democratic party organization in Denver is one of the most spectacular in the country. In 1962–3 the party controlled partisan nominations in the city, was the main organizer of partisan campaigns, and was at the centre of campaigning in mayoral elections. Fifteen years later it was much less important than the candidate in partisan nominations, had a relatively small role in campaigns for partisan offices, and was involved in only a minor way in mayoral and other municipal elections. Of course, the circumstances in which this decline occurred were unusual – the abolition of a very large multi-member legislative district and the failure to convert city elections into legally partisan contests were factors which were especially conducive to party collapse. But it would be a mistake to see the Denver case as being *sui generis*. Legislative and mayoral candidates took over from the party because, like candidates elsewhere, they now had access to campaign resources which were not available to their predecessors twenty or thirty years earlier. As in New York, the particular circumstances affected the rate of party decline and the form which the new candidate-party relations took, but there were also more general forces in the American polity which were weakening the parties. The history of American party decline everywhere involves the interaction of these forces and those specific to a locale. This point must be emphasized before we turn to consider what might seem to be the even more unusual case of the East Bay.

## 3. East Bay: Political Extremism and the Lack of an Organizational Structure

In the 1950s the new extra-legal Democratic organizations altered the role of the party in California as a campaign vehicle and helped to make the party competitive state-wide. It was not an era of organizational decline but of renaissance. Rather like the canals built in early nineteenth-century England, however, these structures were soon overtaken by technological innovation. The canals did not fall into decay – they merely became less efficient than railways – and the same was true of the CDC and its affiliates in relation to the new campaign firms. The weakness of the CDC, such that it could not provide any opposition to its rapidly emerging competitors, was twofold. It was not part of the legally-defined party, and the use it could make of that party was severely limited by the latter's non-existence at the ward or precinct level. If the absence of a formal party structure accelerated the demise of party activity in the changed conditions of the mid-1960s, there was an additional factor in the East Bay which further reduced the role of the Democratic party. This was the growth of popularly supported 'fringe' political movements. In Oakland, where the Black Panther party was formed in 1966, it limited the growth of Democratic-party-centred activities. In Berkeley the impact of a city-wide radical movement was, if anything, greater than that of the Panthers in Oakland, and this movement helped to create a competitive neo-party system at the city level of politics which replaced the earlier Democratic–Republican system. In this section we examine each of these developments in turn.

In spite of its lack of powers, and the fact that there was only limited 'grassroots' representation on it, the Alameda County Democratic Central Committee had developed two functions. The more important function concerned its involvement in Oakland city elections, an intervention made possible by its power of endorsement, while a minor function had been to co-ordinate the production of Democratic 'slate' literature in partisan campaigns.

Although they are debarred from endorsing candidates in partisan primaries, political parties in California do not suffer from this legal restriction in non-partisan elections. As we noted in Chapter 3, Eugene Lee pointed to the endorsements made by the Central Committee in the 1959 Oakland city elections and the campaign effort they made on behalf of a Democratic slate. But the Central Committee was an organization which endorsed candidates who had already decided to run for election – it did not actively recruit candidates. Since the Committee's membership had no mass constituency, it lacked the prestige to be an effective recruiter of political talent. Moreover, the elected public officials who were the Committee's ex-officio members had little interest in sharing their individual influence in the recruiting process with others. The result was that the formal party organization could act only when it was approached by candidates who sought the 'weight' a party endorsement supposedly carried among some voters. Given this restricted form of party involvement, it was not surprising that a policy of systematic endorsement did not provide an electoral breakthrough for the Democrats in Oakland city elections in the 1960s. In response to this failure, a group of Democrats in the early 1970s decided to build on recent successes in partisan elections by having greater party involvement in the nomination of non-partisan candidates in Oakland. In January 1971 they organized an informal convention which any registered Democratic voter was eligible to attend. Would-be candidates presented themselves to the meeting which later voted on the composition of the endorsed slate. The advantage of this procedure was that it attracted, as both candidates and participants, community activists and others who had not been much involved in partisan election campaigns. In doing so, it visibly expanded the Democratic coalition and the party's co-ordinating role. Formal endorsement by the Central Committee followed the convention endorsement, and slate literature was distributed under its auspices. One of the slate candidates was elected to the council that year. Even this, relatively small, success was an advance on previous years and the scheme might have served as a model for future party intervention in city elections but for a fundamental flaw in it; this was exposed by the entry of the Black Panthers into Oakland electoral politics.

In 1973 Bobby Seale, the radical black activist and former defendant in what was originally the 'Chicago Eight' trial, decided to run as the candidate of the Panthers in the Oakland mayoral election.[19] One effect of this was to split the alliance between white liberals and blacks which had been organized around the Democratic slate two years earlier. In these circumstances the question of whether a convention should be arranged never arose, and Republican control of the city was not threatened. The Panther connection scared many Democrats, including some prominent liberals, and it was against this unfavourable background that a number of East Bay elected public officials attempted to revive the informal nominating convention in 1975. The crucial problem was that at least one of them, congressman Pete Stark, refused to countenance a proposal for a convention which any registered Democrat could attend, from fear that the meeting might be packed by

Panthers. On the other hand, in the absence of a formal precinct organization, there was no obvious basis for constituting a more select meeting with activists who might be expected to have some loyalty to the Democratic party rather than to the Panthers. Even after extended discussions, no convention could be convened, and the candidate elected in 1971 remained the only liberal Democrat on the city council. The failure of the Democratic party in Oakland over the previous quarter of a century is apparent when we realize that by 1975 there were very few Republicans serving as elected public officials in the East Bay, except on that city's council, and that whites were already a minority of the city's population.

In 1977 the era of electoral failure ended. Instead of reverting to the kind of convention used in 1971, however, the major elected Democratic politicians in the area negotiated informally among themselves and unanimously agreed to support a black judge, Lionel Wilson, as the mayoral candidate. These negotiations did not produce a slate of candidates, but the Central Committee did endorse a slate consisting of Wilson and two candidates for the council who had announced their candidacies independently. Subsequently, campaign literature referred to the three candidates as the 'official Democratic slate'. Because of this they were widely seen as a team of candidates, although they campaigned as individuals and not as a team. Of course, the earlier failures by Democrats cannot be attributed solely to the consequences of weak party structures. The low visibility of the council offices, low turnouts by the black electorate, and the rise of the Panthers in the second half of the 1960s were, undoubtedly, more important factors. Nevertheless, the fact that the party's Central Committee could act as an endorsing agent only when approached by candidates restricted the party's role in candidate recruitment. Even without the problem of the Panthers, the convention device used in 1971 had the disadvantage that it was susceptible to 'packing' by individual candidates, and this problem would have become worse as Democrats became more competitive in the city. Indeed, by 1979 the party's improved competitive position actually helped to prevent the Central Committee from playing effectively its now established role of endorsement.

There were two causes for this breakdown in the committee's role. First, the small, and unrepresentative, committee endorsed one of its own members in a council district rather than a much stronger Democratic candidate. This removed the credibility of its slate. Secondly, in another district a bitter contest emerged into which most of the area's elected officials were eventually drawn – a campaign which separated liberal–radical from moderate Democrats. The result was that the one remaining (and predominantly) white Democratic club in the city (the Montclair-Greater Oakland Democratic Club, or MGO) joined the black Niagara club in endorsing a rather different slate from that of the Central Committee; the two clubs then distributed their own slate literature. The MGO had always put out its own literature, but this was the first time it had either co-ordinated with another club or endorsed a slate which differed from the official one. Thus, in Oakland there was a rapid switch from a city where the Democratic party was unable to mobilize effectively for municipal elections to one where the factional politics of a one-party area rendered the formal party organization redundant.

The devaluation of the Central Committee endorsement in 1979 meant that some slate members refused to pay the usual contribution for the slate's campaign literature (or 'mailers'). Without this income the Committee's mailer could not be produced, and it was this mailer which, drawing on party loyalty among Democratic voters, had always been the valuable element in the party's endorsement. In partisan elections the Central Committee had always put out similar slate literature but this had been of importance only to lesser office candidates, for whom any emphasis on straight-ticket voting helped in what had become a one-party area. Normally slate literature involved only the official party slate, whether it was put out by the Committee or the MGO, but there was one occasion in 1976 when four candidates in the East Bay produced joint billboards. Although it was not described by them as publicizing a slate it was in effect doing just that.[20] Two of the candidates, congressman Dellums and assemblyman Bates, had little to gain from the venture, and its main beneficiaries were political allies running for non-partisan offices, both of whom had Democratic opponents at the general election. Unless, as in this case, a major office candidate had political debts to settle, there was virtually no incentive for him to participate in such ventures: he would have to be a disproportionately large contributor to the costs, and yet could expect to gain few votes from it. Such candidates only contributed to the literature for official slates out of a sense of party loyalty; the Central Committee was never in a position, as the Denver party had been before the 1960s, to demand financial contributions. By the end of the 1970s, though, with the establishment of factions centred on elected public officials, neither the Central Committee nor the other party organizations (such as the MGO) had any significant role to play in Oakland elections. This represented a transformation from its position in the late 1950s. This conclusion about the obsolescence of party organization in the East Bay becomes even more apparent when we turn to examine the growth of a neo-party system in Berkeley in the 1970s.

With the election of a Democrat in what was now called the 16th Assembly District (the old 18th District) in 1970, the Republicans lost their last major partisan office in the northern East Bay and, worried by radicalism in the 1973 Berkeley city elections, Berkeley Republicans withdrew in favour of moderate Democrats. In the absence of serious Republican opposition, partisan politics *per se* could not be practised in the city. Yet politics there was to remain fiercely competitive, not only with respect to public offices, but also over many public issues. As we shall now show, that competition cannot really be characterized as being between two *factions* of a party, because in an area with voter registration overwhelmingly for one party and with loose party structures, the competition did not extend beyond city politics. Very little of importance to the parties was dependent on what happened in Berkeley local government, and Berkeley was able to develop organized electoral competition between two groups divided by political ideology, neither of which was an identifiable faction of a party. Perhaps the best characterization of this is that Berkeley had party-type institutions (or neo-parties) at the city level of politics, which were independent of the parties themselves.

A realignment of political forces in the city occurred between 1969 and 1973. 1971 was the turning point. That year radical-supported candidates won three of the four council seats because moderate Democrat and Republican candidates split the non-radical vote. The white radicals employed a tactic similar to that used by white Democrats in the 1950s – they instituted a coalition arrangement with a black political organization, in which a high degree of autonomy was left to the latter. However, the so-called Black Caucus was a body with no formal membership, and its endorsement meeting was 'stacked' by political irregulars; the two candidates whom the Caucus nominated were so disruptive once in office that they discredited radical politics. From the experience of this 1971 election both the radicals and the non-radicals drew the conclusion that much 'tighter' political organization was necessary for long-term electoral success.

The non-radicals moved first. Republicans withdrew from city election contests, thereby reducing the number of candidates, and then moderate Democrats arranged an informal meeting to endorse a slate of four candidates who would run as a team in the 1973 elections.[21] This was almost pure caucus-cadre-style politics – a group of like-minded political activists were selecting the candidates. It was remarkably successful in imposing discipline on the electoral process. In 1973 only one candidate rejected by the meeting broke rank and continued his candidacy at the election. By 1975 responsibility for selecting candidates had been transferred formally to the Berkeley Democratic Club (BDC), with its members voting on endorsements. This may have given the candidates greater legitimacy among a few voters, but this was the only real difference from the 1973 method of selection. Co-operation over nominations and campaigning by the BDC candidates worked: between 1973 and 1979 they controlled the city council. In many respects the BDC might appear to have undertaken activities normally associated with a political party. Its relation to the Democratic party in the state and in Alameda county, however, suggests that it is misleading to view it as a unit of the party, as a faction of the party, or as the successor to the Berkeley Caucus.

Those who wished to argue that the BDC really was a faction of the Democratic party could point to several seemingly partisan features. Nominations for a set of candidates had come to be controlled by a political club affiliated to the Democratic party. Most of those active in the nomination process were Democratic voters. Many of the BDC candidates were endorsed by the county Central Committee and received endorsements from Democrats elected to partisan public offices. Superficially, this might seem similar to party involvement in Berkeley elections between 1953 and 1965. Yet we would suggest that this interpretation should be rejected. There are two main reasons for this. One argument is that the BDC was a very different body from the clubs of the 1950s or the Berkeley Caucus. It was little more than a label; it simply acted as the vehicle for nominating a slate of candidates. The BDC in the 1970s was really a bipartisan institution: it sought, and obtained, financial contributions and votes from Republicans, and its candidates were regularly backed by the Republican-oriented daily newspaper, the *Berkeley Gazette*. In brief, there was no party-oriented electoral mobilization promoted by the BDC. Unlike the Berkeley Caucus, it was not one arm of a wider movement which was concerned with organizing nominations and campaigns for partisan offices. A second reason relates to the internal operations of parties. By the mid-1970s there was virtually no activity or institution of any importance in the East Bay which could be said to be a party activity or institution. For something to be a party faction there have to be party positions to be captured and activities to be organized, as well as opponents in the party who want to control these positions and activities themselves. The BDC had opponents but it inhabited a political universe in which the personal organizations of candidates could perform all the tasks for which party factions might be needed. The BDC was not merely a rather inactive party faction, it was a body which could not have acted as a faction even if it wanted to, because the processes it might have controlled had been assimiliated by candidates and other more specialized bodies. It was a participant in local electoral politics, but not at other levels; for this reason, we would argue that it is best seen as a neo-party.

The growth of a neo-party linked to, but separate from, the Democratic party is also evident in Berkeley's radical movement. The formal coalition between the white radical group and the Black Caucus was abandoned after the disastrous consequences of the 1971 elections became apparent. From then on, the process of building links with the black electorate was conducted through personal contacts with black political and community leaders. The radicals also gave increased emphasis to problems at the neighbourhood level, and the result was a fusing of the nationally-oriented radicalism of the 1960s with the newly emerging community activist movements. The formal link between these two movements was initiated in 1974 when a meeting of about fifty representatives was convened to launch Berkeley Citizens Action (BCA).[22] Under this label a slate of candidates who identified themselves as 'progressive', rather than 'radical', contested the 1975, 1977, and 1979 elections.

At first BCA was a loose coalition of groups and individuals, but after defeat in the 1977 elections it established a formal membership and was then partly financed by the dues paid by individual members. This was one of the changes which made it possible for the group to employ staff and have a permanent headquarters. The other important change was the birth of the Campaign for Economic Democracy led by Tom Hayden, of which the BCA became the Berkeley affiliate. New organization and finance produced a BCA campaign in 1979 which was radically different from previous years – one which gave much greater emphasis to mailers at the expense of door-to-door campaigning. If in this new form the BCA seemed to resemble an embryonic mass-membership political party, there are two reasons for disregarding this. In the first place, a large number of members, possibly as many as half the total, were vehemently opposed to BCA participation in electoral politics outside Berkeley city elections. For them the purpose of the organization was community mobilization, not electoral politics, and many of them were antagonistic to the Democratic party. To preserve its internal cohesion, the BCA had no alternative but to restrict its electoral role and its contacts with the Democratic party. The second reason is that by 1979 the BCA had become more 'permissive' in its relations with candidates to whom it gave an endorsement. In earlier years there had been little individual campaigning by candidates, but in 1979 they were actually encouraged to mount their own efforts in addition to those of the slate. Behind this strategy lay the assumption that crucial extra votes might be gained by a candidate who could appeal to people who would not normally vote for the BCA slate.[23] In doing this, the BCA was moving away from the disciplined, team approach to the type of campaigning usually associated with mass-membership parties.

Even more than the BDC, the BCA had close relationships with some public officials outside Berkeley local government. The closest relationships were with congressman Ron Dellums and county supervisor John George, both of whom acted as important links with the black community. It is necessary to emphasize, however, that the connections with Dellums, George, and with white assemblyman Tom Bates were with individual elected officials, and not with any entity which could be described as the Democratic party or as a faction of that party. Dellums, George, and Bates were not a party faction, even though they were sometimes described as the 'East Bay Machine'.[24] They owed their power solely to the public offices they occupied, and not to any party structures. This point is most apparent if we compare Bates's relationship with the other two with his predecessor's relationship with Dellums. While Ken Meade had radical sympathies, rarely did he work closely with Dellums either in campaigning or in any other respect; he viewed their interests as distinct because he had a largely white Assembly district, while Dellums's electoral support was largely black.[25] Bates took a different view of co-operation when he succeeded Meade. He saw mutual advantages in a limited number of shared electoral activities for those who had the same political philosophy. These included the informal slate posters in 1976 and some co-ordination over endorsements of other candidates, so that embarrassing schisms could be avoided. But this was the limit of the integration of electoral activities; Dellums could no more protect Bates electorally than Bates could add weight to Dellums's campaigns.

We have devoted considerable space to explaining what happened to politics in Berkeley. The point of doing so was not that Berkeley was a typical American city; it was not, and in many ways it was one of the most unusual cities in the country. The highly-organized neo-parties at the city level of politics were certainly exceptional. The point of devoting some attention to the city is that the Berkeley example exposes in an exaggerated form some general problems facing Democratic parties in American cities. In the same way as many others, it had become a one-party area by the late 1960s, so that opposition to a common enemy did not tie its former constitutent groups to the party. The lack of a formal, legally-defined party structure made this problem much worse, so that a new challenger, the radical movement, could destroy the old Democratic club movement. But the most revealing feature is that the rise in importance of neighbourhood issues did not lead to a re-establishment of party politics in the 1970s. Parties were unnecessary for this.

Berkeley was far from unique in experiencing an upsurge of popular concern about the conditions of neighbourhoods; neighbourhood associations and action groups expanded in many cities. Indirectly, these issues were tied to others, such as resentment of rising property taxes and rent control. Yet, even in the Berkeley case, where the Democratic party had been well-placed to latch on to these issues and to operate in the way that Liberal parties do in some British cities, the party was bypassed in the growth of neighbourhood politics. The failure of American parties to regroup around local issues is, perhaps, as significant for their decline as the rise of the individual candidate in electoral politics. Yet it is not surprising, for the caucus-cadre party is ill-suited for this. Despite the fact that such parties are locally based, they are organized around the winning of offices; it was this which gave the Berkeley club movement its momentum. In some circumstances issue mobilization can make a contribution to this end, but it does not constitute part of that objective. The caucus-cadre party is not adaptable to political activity which is focused on objectives other than winning certain offices. When its primary objective becomes obscured or unrealizable, as it was in Berkeley and Oakland because of political extremism, there is no other source of cohesion and collective action to which the party can retreat. When there is no formal–legal structure for its organizations they simply dissolve, as they did throughout the East Bay. Even with a legal structure, however, the problem for the caucus-cadre party is to maintain a belief among potential activists that participation is still worth while, even though temporarily 'office-hunting' has become difficult. This problem is hard to resolve precisely because such a party has organized itself around the pursuit of public offices. What was different about the period 1960–80 was that, by the time normal politics was resuming in the mid-1970s, there were instruments other than parties for seeking control of public offices.

In this chapter, we have tried to explain how party structures and mechanisms became weaker in each of the three areas, and how attempts to revive the role of the organizations failed. We have suggested that rather different combinations of factors were partly responsible for this in what were three very different kinds of party organization. If the study was replicated elsewhere, further patterns of structural collapse would emerge. The important point, however, is that throughout America disloyalty to, and 'exit' from, party structures was becoming possible in a way that it had not been before, due mainly to the growing independence of individual elected politicians. Internal party disputes were constrained no longer by the institutional frameworks of the 1940s and 1950s, when there were fewer alternatives to remaining in the party or joining another one. When normal politics returned, and when issue-oriented activists were once again committed to their party, they faced a world in which candidates and single-issue groups were formidable rivals for the time of occasional activists.[26] By the late 1970s the parties could not recruit effective organization members, and the organizations' powers were that much less than they had been twenty years earlier. They had lost out to a 'new notability' – the candidates.

## Notes

[1] The literature on these reforms, and their aftermath, is vast. Among the more important works are William J. Crotty, *Decision for the Democrats*, (Baltimore: Johns Hopkins Press, 1978); James Ceaser, *Presidential Selection* (Princeton: Princeton University Press, 1979); Nelson W. Polsby, *Consequences of Party Reform* (Oxford and New York: Oxford University Press, 1983); and Byron E. Shafer, *Quiet Revolution* (New York: Russell Sage Foundation, 1983).

[2] *New York Times*, 12 Feb. 1976.

[3] Ibid., 13 Mar. 1977.

[4] Ibid., 9 July 1978.

[5] Ibid., 3 Aug. 1982.

[6] An example of this claim about Brooklyn is the long article in the *New York Times*, 10 Dec. 1972. In spite of the party's reputation, neither leader Meade Esposito's funds nor his party staff rivalled that of the Republican machine in nearby Nassau County. In 1974 it was reported that Esposito's staff consisted of three people and the party's annual income was about $100,000; the Republicans in Nassau County had a twenty-three-person staff and an annual income of about one million dollars. The large party income there was the result of 'macing' – a 1 per cent charge on the income of all county employees. *New York Times*, 7 July 1974.

[7] Nelson W. Polsby, 'Mayor Daley and the Bishop', *Political Studies*, 28 (1980), 465.

[8] *New York Times*, 7 Nov. 1971.

[9] For the background to Troy's attempt, see *New York Times*, 3 June 1974.

[10] *New York Times*, 9 Nov. 1974.

[11] *New York Times*, 3 July 1980.

[12] Howard A. Scarrow, *Parties, Elections and Representation in the State of New York* (New York and London: New York University Press, 1983), p. 46.

[13] *New York Times*, 16 Sep. 1965.

[14] Maurice Klain, 'A New Look at the Constituencies', *American Political Science Review*, 49 (1955), 1005–19.

[15] This expression, of course, derives from the work of E. E. Schattschneider.

[16] Austin Ranney, 'Parties in State Politics', in Herbert Jacob and Kenneth N. Vines (eds), *Politics in the American States*, 1st edn. (Boston: Little, Brown, 1965).

[17] In *Baker v. Carr* (1962).

[18] Ware, *The Logic of Party Democracy*, pp. 102–6.

[19] Seale was eventually tried separately, and the trial was henceforth known as the trial of the 'Chicago Seven'.

[20] Interview nos. 20 and 21.

[21] Interview no. 34.

[22] Interview no. 51.

[23] Interview no. 51.

[24] Articles in both the *Oakland Tribune* and the *California Journal* would often refer to 'Dellums's East Bay Machine'.

[25] Interview no. 7.

[26] On the commitment of party activists to their parties at 1980 state conventions see Abramowitz, McGlennon and Rapoport, 'The Party Isn't Over: Incentives for Activism in the 1980 Presidential Nominating Campaign'.

# 6
# The Rise of the New Notability: Candidates

The relationship between a party candidate for public office and his party has changed considerably since the days of Hymie Shorenstein. A district leader in the Brooklyn Democratic party in the 1920s, Shorenstein provided one of the best descriptions of two types of party candidate (the person at the head of the ticket and every other candidate) which were to be found in most states.

> ... [he] was once confronted by a worried Assembly candidate who had not received the funds he needed to campaign actively. Shorenstein calmed him: 'You see the ferryboats come in?' Shorenstein asked. 'You see them pull into the slip? You see the water suck in behind? And when the water sucks in behind the ferryboat, all kinds of dirty garbage comes smack into the slip with the ferryboat? Go home. Al Smith is the ferryboat. This year you're the garbage.'[1]

Until about 1960 what might be called the Shorenstein 'dirty garbage' model of campaigning (more popularly known as the 'coattails effect') provided a fairly good account of the role of most candidates for lesser offices at general elections this century. With straight-ticket voting, the fate of these candidates depended on two factors: the marginality of the particular electoral district, and the performance of the candidate at the head of the ticket in conjunction with the effort of the party organization. In a presidential year it was the party's presidential hopeful who was the 'ship' to suck in the 'dirty garbage'. In a mid-term election this role would normally be performed by the gubernatorial candidate, though to a lesser extent the US Senate candidate might help the ticket.

Of course, there are several qualifications to be added to this account of the applicability of the 'dirty garbage' model. First, what Schattschneider called the 'system of 1896', the party system which lasted from the mid-1890s until the early 1930s, was characterized in most states by the dominance of one party.[2] In these states even a disastrous candidate at the head of the ticket might not harm greatly the election chances of the party's candidates for lesser offices. Secondly, in some years the mid-term elections of a few states (normally about five until the 1940s) would not feature an election for the governorship or the US Senate, and in such cases the 'sucking power' of the major candidate (a congressional candidate) was considerably less than in the Shorenstein model. Thirdly, while the head of the ticket could 'suck in' garbage at the bottom, its effect on middle-range offices, such as congressional seats, was less. Constituency influence on the division of the vote was high in nineteenth-century congressional elections and, although it declined with the nationalization of American politics in the twentieth century, congressional seats remained subject to non-national forces. Even in the 1950s, when constituency and state influences were relatively weaker than they had been earlier, 22 per cent of the seats which 'changed hands' between the parties did so *against* the national tide.[3] Fourthly, in marginal districts for lesser offices campaigning could make a difference. As we have already noted, state legislature candidates in Denver in the 1950s did undertake some campaigning for themselves, in spite of the large size of their electorate. Because there was always some split-ticket voting, an inactive candidate could be defeated while his fellows on the ticket were being elected. Finally, in places such as California, where campaigning was organized around the office rather than the party – a political style made possible partly by the cross-filing primary – the 'dirty garbage' model had much less relevance. Nevertheless, even allowing for these qualifications, it is undeniable that in many American cities the 'pull' of candidates for major offices substantially reduced the amount of campaigning other candidates had to undertake on their own behalf.

This 'pulling power' of the party might seem rather odd, however, as American parties have remained decentralized structures since their formation. In brief, it might be asked, why was there any 'pull' there? Leaving aside the forces which nationalized American politics during this century, we can identify three features of late nineteenth-century politics which contributed to party 'pull'. First, in many places there were highly organized structures, usually at the county level, that had an interest in mobilizing the vote as efficiently as possible. Secondly, after the adoption of the secret ballot, these machines lost a major source of *local* control over the electorate; more indirect means of structuring the vote than many of those used previously had to be relied on. There was a greater incentive to draw on voter attraction to 'big name' candidates, and this meant linking local campaign efforts more to state and national efforts. Thirdly, the frequency of elections meant that party campaigns were advantageous to candidates. (Many governorships, for example, were contested biennially.) They helped to simplify the choice for the electorate and reduced the resources that would have to be expended in winning a particular office. In the early twentieth century, therefore, the 'dirty garbage' model of campaigning was attractive to both parties and candidates. In the 1930s Franklin Roosevelt's 'national' solution to the economic depression, and his use of the new medium of radio to campaign for this and for himself, further increased the incentive to make use of party 'pull'. Perhaps 'dirty garbage' was the model for the future.

What prevented the further development of party-centred campaigning was not a reversal of the nationalizing tendencies, but a change in the conditions which made the 'dirty garbage' possible and further changes in campaign technology. The urban party machines were slowly decaying; the terms of state offices were lengthened, and many gubernatorial elections were no longer contested in presidential years; campaign technologies which at first could be used only by major office candidates were gradually employed by other candidates, and these were methods of campaigning which were not controlled by party organizations. The 1950s was the last decade in which a plausible case could be made for a candidate to use 'dirty garbage' politics. From then on, 'coattails effects' became smaller as voters' propensity to ticket-splitting increased. This encouraged candidates to campaign more independently of their parties and this again reduced voter attachment to the parties.

In this chapter we are concerned with three issues relating to the decline of the 'dirty garbage' model. In the first section we consider briefly the evidence which supports the commonly held view that few candidates are now at the mercy of party 'ships'. In the second section we examine how legislators of all kinds have transformed the jobs they do so as to put themselves in the best possible position to withstand party 'tides'. Finally, we outline a rather unexpected development, that the growing independence of public office-holders is not tending to produce a uniform system of atomized electoral competition in America. On the contrary, there are circumstances in which elected public officials have an incentive to intervene actively in the campaigns of others, even when there has been no tradition of this in their state.

## 1. The Declining 'Pull' of Parties

Ticket-splitting occurs when voters have knowledge of a candidate (whether it be his style, the issues he supports, or simply his name) which is independent of the knowledge they have about his party. The candidates who are most capable of divorcing themselves from the party campaign effort are those seeking offices of such value that it is rational for them to spend considerable resources. The less important the office, the less central it is for the careers of the politically ambitious, and the less that is known about the office by the voter, the smaller will be the amount of electoral resources it is worth expending in the campaign. Thus, some very minor offices will simply reflect the distribution of party voting at an election; generally, they will be unaffected by the individual efforts of the candidates, unless an individual is prepared to commit resources vastly in excess of what is usually considered to be the 'value' of the office. Consider the case of the Regents of the University of Colorado. Each of them is elected on a partisan ballot in a state-wide election. The position has never been a paid, full-time job, nor is it a recognized stepping-stone to other political offices, and it has extremely low visibility in the state. While Regent candidates do distribute some of their own campaign literature, it is not worth while for them to distribute much, and it is certainly not an office which would attract extensive media campaigning. Hence, the office has remained largely unaffected by ticket-splitting, and it is used by some political experts as the best measure of the division of the party vote in the state.[4] In Shorenstein's language, it is 'garbage' but today it is not sucked in by the major office candidates, but is subject more to the forces that affect the division of party loyalties in the state; the most appropriate analogy, perhaps, is that it is influenced by tides and currents rather than by a ship. Yet, and this is our central point, to obtain a good, contemporary example of the 'dirty garbage' phenomenon we have had to turn to a particularly obscure office. An observer transported forward in time from the late 1940s would probably be most surprised by the extent to which congressional and state legislature candidates, as well as candidates for state-wide office such as attorneys-general, have been largely freed from the 'sucking in'.

As every student of American electoral behaviour knows, ticket-splitting among the electorate increased significantly between the early 1950s and the early 1970s, with the Nixon–McGovern contest of 1972 generating the greatest ever number of split victories. In presidential election years between 1920 and 1944 ticket-splitting was 'decisive' in no more than 19 per cent of House districts. From 1956 onwards the lowest incidence of 'decisive' ticket-splitting was in 1960 when it was to be found in 20 per cent of districts; in 1972 it occurred in 44 per cent of congressional districts.[5] It was not until the early 1970s, however, that ticket-splitting at the state level was regularly producing divided control of state government (see Chapter 3). Together with the increase in party competition in many states, the willingness to split the ticket reduced the relevance of the 'dirty garbage' model. The reduced 'pull' of the parties can be seen in many ways. There is the evidence of voter surveys about ticket-splitting behaviour.[6] There is also the evidence of the low level of incumbent defeats in general elections: in both the Congress and state legislatures it has rarely been above 10 per cent of those seeking re-election, and often less than 5 per cent in recent years.[7] We can also point to the increasing number of landslide victories in congressional contests, something which is found in state legislature elections as well. Nevertheless, in citing this evidence, it is important that we do not misunderstand the relevance of the increase in the number of safe districts.

One mistake which might be made is to assume that the extent of party 'pull' can be determined by examining the number of districts changing hands between parties. If this were so, we should have to admit that party 'pull' declined greatly from the late nineteenth century to the 1950s. In fact, what happened was that fewer districts remained two-party competitive but, in those districts which were competitive, the candidates became exposed increasingly to the effects of nationalizing tendencies. Most congressmen were not 'garbage' dependent on the effort and the appeal of the party's more visible candidates. The minority of them who were heavily dependent on the party's 'ship' were the incumbents of the party marginal districts, districts in which there were similar numbers of Democratic and Republican identifiers in the electorate. In these districts an incumbent could expect a relatively short continuous career in Congress. He would be elected during the party's surge in the electoral cycle, and would usually be defeated in the decline phase. In some cases the incumbent's own campaign efforts could prevent this, but in his early years in the House its seniority system frustrated effective self-help. Freshman and second-term congressmen acquired virtually no power within the chamber, and thus could not appeal to voters in terms of their indispensability in protecting district interests. Of course, the power which came from increased seniority aided incumbents in two ways. It could be used by all of them as an electoral issue against primary challengers and, with the obvious exception of those from one-party districts, it was also an important resource with which to counter a general election challenger in a year of 'surge' for the opposing party. Although there would be many exceptions to this generalization, we can say that an incumbent congressman was more likely to be 'dirty garbage' than a 'small ship' in the following conditions: (i) the more evenly divided the partisan support within his district, (ii) the more congressional districts there were in his state, and (iii) the less seniority he had.

When considering 'party pull' it must moreover be emphasized that the distinction between marginal and safe districts has never been as clear in the American Congress as it is, for instance, in the British House of Commons. Candidate unpopularity could turn a supposedly safe district into one in which an incumbent could be defeated or have only a narrow margin of victory. Equally, as we have already seen, a supposedly marginal district might go against the national tide; in the 1950s, when more than one in five of the congressional districts which changed hands did so against the party tide, constituency deviance was about three times as common as in Britain. In such circumstances the candidate's decision as to how closely to link his own campaign with that of the party was a complex one, and it was not connected solely with his belief about the marginality of the district. For example, in a district which was relatively safe, Byron Rogers chose to identify his own campaign with that of the party rather than choosing, as he might have done, to insulate himself from national Republican surges through more independent campaigning. As we have seen, this was rational for him, given the potential ideological split within the city's Democratic party. What evidence there is from this earlier period suggests that there was no typical relationship between the congressional campaign and the party campaign, nor was there any single factor which could explain the variations in this relationship.[8]

But, if there were about thirty party marginals in the Congresses of the 1950s, how many were there by the 1970s? That is, how many congressmen were exposed to shifts in partisan support (were 'garbage-like') rather than being able to control their own electoral destinies? Before suggesting how we might estimate the number of marginal districts, it is useful to bear in mind three well-known facts about congressmen. First, at any given election most of them win re-election;

usually over 90 per cent have done so, and in half the elections between 1960 and 1980 over 95 per cent of them were successful. Secondly, as David Mayhew among others has pointed out, most congressmen have at least one close election during their careers, so that complete electoral security has been enjoyed by only a minority.[9] But, thirdly, the size of incumbents' electoral pluralities was increasing in the 1970s; more congressmen were winning with at least 60 per cent of the vote than did so in earlier years. In the six elections between 1956 and 1966 the proportion of incumbents who obtained 60 per cent or more of the two-party vote varied from 58.5 per cent of the total (1964) to 67.7 per cent (1966). The mean for the six years was 61.8 per cent. In the following seven elections (1966–80) the corresponding figures were 66.4 per cent in the least successful year for incumbents (1974) and 78 per cent in the most successful year (1978). The mean for these seven elections was 73.8 per cent.[10] Thus, what we are trying to assess is the number of *party* marginals given that the typical congressional career displays a pattern of occasional vulnerability amidst general security.

There are a number of methods which might be used in deciding which districts were party marginals in the 1970s. The following four criteria would seem necessary, though they may not be sufficient, in isolating these marginal districts. We have applied them to the period 1972–80, which is the complete period of electoral contestation using the district boundaries drawn up after the 1970 US Census. The four criteria are:

(i) A district must have switched from one party to another and back again at least once during this period to be truly marginal.

A district cannot be held to be marginal if it is unaffected by surges and declines in electoral support for the parties; our period does include a surge-decline cycle – from the slight Republican surge in 1972, through the party's major decline in 1974 and 1976 to a surge for the Republicans in 1980.

(ii) An incumbent must have been defeated in at least one general election.

The reason for including this condition is that it helps to eliminate the possibility that switches in party control of a district were the result solely of an imbalance in the strength of the candidates selected, rather than being at least partly related to partisan forces. When, for example, two switches of control involved a retirement and the new incumbent running for another office, it is probable that, in an age of split-ticket voting, these changes would not reflect the surge-decline cycle. The disadvantage of this condition is that it is too strong; it will lead to an underestimation of the number of party marginals, because there may be some districts where the incumbent chose to retire, or run for another office, rather than facing electoral defeat. We might expect, though, that the total number of such districts would be much fewer than the number of marginals we otherwise identify.

(iii) The exclusion of instances in which the switch of party control was directly attributable to unusual circumstances in a special election, to a political scandal involving the incumbent, or to court-ordered reapportionment in California.

This condition prevents the inclusion of 'freak' results which might otherwise inflate the marginal district category.

(iv) The exclusion of instances involving candidates winning with more than 60 per cent of the vote in any of the five elections.

When a candidate has been able to win 60 per cent of the vote, the district ceases to be marginal, for he has been able to develop his personal support to a level which places him outside the effective range of partisan forces. Of course, this is not to say that a congressman who obtained this share of the vote was safe at the next election. Inattention to his district, demographic changes, and scandal, might all lead to defeat; indeed, of the twenty-seven incumbents who lost in general elections in 1980, fourteen had won 60 per cent of the vote in 1978.[11] The point is that changes in partisan forces do not account for switches of this kind.

When we apply these four criteria, we find that only five districts in the country can be considered to be party marginals in the 1970s: Michigan Sixth, New Jersey Seventh, New York Third, and Virginia Eighth and Tenth. Even if we allow for the caveat indicated under (ii), the number of marginal districts is unlikely to be more than 2 per cent of all districts. On their own, party fortunes have relatively little impact on the careers of congressmen. The relatively small minority of seats which were 'garbage' thirty years before had shrunk to a tiny minority in the 1970s. Of course, this does not mean that, in all but about five districts, congressional elections take place in a context divorced from national politics or opinions. Far from it, and congressional elections are linked to these currents through the quality of challengers selected and the funds and other resources channelled to them.[12] However, the modern congressman has far more resources than his predecessors with which to withstand tides for the opposing party, and it was this fact which enabled so many of the 1974 freshman members to stay in office throughout the 1970s.

When looking at congressional seats it has been possible to indicate fairly precisely the decline of party 'pull'. However, when we turn to consider the offices at the centre of the 'dirty garbage' model, the lesser state offices, state legislature seats, and so on, we find this is not possible. There have been very few studies of turnover in state legislatures and none on the impact of party 'pull' on this turnover. Turnover has been much higher than in Congress, but it declined steadily from the 1930s to the 1970s; in eight states studied by Jewell, general election defeats for incumbents in the 1970s occurred about as frequently as they did in congressional elections.[13] This would indicate that party 'pull', perhaps, is now no more effective in state legislature elections than in congressional elections. But what of its effect in the 1950s, or in Hymie Shorenstein's time? There is cause for believing that Shorenstein was correct – the party 'ship' did suck in the 'dirty garbage' – but there are no data on this. The reasons for believing that the party had much greater 'pull' over lesser offices than in congressional elections then, and much greater 'pull' then than now, include the following. Many of these offices were part-time, low-paid positions for which there was little incentive to engage in much individual campaigning; the greater use of heterogeneous, multi-member districts in the pre-reapportionment era; and the frequency of elections (New York, for example, had annual Assembly elections until 1936) encouraged party-centred campaigning. By the 1970s state assembly contests in some states had become more like congressional contests than they had been in the 1930s or the 1950s: the legislatures were more professional and provided a greater incentive for the incumbent to preserve his political career; some of the new campaign techniques enabled many legislators to campaign largely on their own; there were fewer multi-member districts; and there were no annual Assemblies. Circumstantial evidence, therefore, strongly suggests that there has been a major reduction in party 'pull' in lesser office contests.

## 2. Increasing the Resources for Incumbent Independence

Incumbent legislators have not merely been the beneficiaries of the revolution in campaign technology; they have, in many cases, altered their conditions of service in ways which indirectly improved their own electoral prospects. This has weakened whatever dependence they may have had on their party organization. Not only congressmen, but also state legislators in some of the larger states, have been able to transform their jobs so that in one sense they can be said to be campaigning permanently. We begin by examining Congress, for this has served as a model for what has happened at state level.

Before the 1950s there were obvious constraints on the ability of a congressman to campaign effectively on his own, rather than in conjunction with his party.[14] One of them was the limited range of techniques available with which to 'reach' the electorate. With the number of voters he faced, personal contacts and friends could not provide the extensive network of information and propaganda that even a decaying party precinct organization could. In the absence of sophisticated opinion polling, there was no other guide to the mood of people who, for most of his term of office, were separated from him by hundreds or thousands of miles. Radio, while useful in the 1930s and 1940s in helping to improve public awareness of the names of some congressmen, could not be employed in the same way as television to develop an image of the incumbent's political style. Again, while campaign mail could be sent directly to voters, it was an inflexible form of campaigning, because there were no devices for effective targetting or for discovering the patterns of response to this literature. If the revolution in campaign technology was, at least initially, of less use to the congressman than to the US Senator or state governor, it did increase the strategies available to the former when deciding how to link his own election efforts with those of the party.

The choice of strategies before the 1950s was limited by the remoteness of Washington from many congressional districts, and this helped to make 'continual' campaigning impossible. Train or car journeys to the west coast took several days, and it was only after the Second World War that there was a comprehensive long-distance airline service which placed the entire continental United States within a few hours reach of the capital. Even when air travel to the districts had become a possibility, Congress did not immediately provide members with the funds to make frequent trips home. Nor did it distribute funds for large staff offices in the districts. George Miller, first elected in his East Bay district in 1944, was not unusual among congressmen from distant constituencies, and as a Californian he was among those who could expect the least help from a party organization in maintaining links with his electorate. In 1945 he was allotted $9,500 for 'clerk hire' and the maximum permissible size of his staff was five.[15] In fact, he employed only one aide, his wife, and he had no district office, only an arrangement whereby mail would be sent on to him in Washington from the district. There was no allowance then for telephone calls from his Washington office to the district, which usually he could visit only once a year – during the summer recess. He would drive out to California with his family in July, although in election years this trip was made in April, so that they could organize the campaign for the June primary, with Miller himself returning later to Washington for the end of the congressional session.[16]

The growth in resources available to congressmen for district liaison was spectacular. In 1947 the typical congressman had three aides (see Table 6.1), nearly all of whom were based in Washington. By the end of the 1970s his modern counterpart had a paid staff of sixteen, though some congressmen did not employ as many staffers as they could. From 1975 they were entitled to employ up to eighteen staff members each, with their combined salaries to be no more than $238,000 per congressman. The most significant change, though, was in the number of staffers attached to district offices rather than Washington offices. The proportion of district staffers increased from 22.5 per cent to 34.6 per cent of the total number of staff members between 1972 and 1979 (see Table 6.2). Together with the increase in staff, congressmen have voted themselves free travel facilities to their districts, as well as free mail and telephone facilities, and they have increased these allowances regularly. Of course, this does not mean that all congressmen have chosen to place equal emphasis on contact with, and service for, their districts when planning future campaign strategy. As Richard Fenno has pointed out, there are a variety of 'home styles' which may be adopted, not one dominant model.[17] Yet, even those who de-emphasize constituency service still visit their districts more frequently,

**Table 6.1**: Personal staffs of Members of the House of Representatives

| | Total number of staff members | Mean number of staff members per congressman |
|---|---|---|
| 1947 | 1440 | 3.3 |
| 1957 | 2441 | 5.6 |
| 1967 | 4055 | 9.3 |
| 1977 | 6942 | 15.9 |
| 1979 | 7067 | 16.2 |

*Source:* John F. Bibby, Thomas E. Mann and Norman J. Ornstein, *Vital Statistics on Congress 1980*, Washington, DC: American Enterprise Institute, 1980, p. 69.

**Table 6.2**: House staffs based in District Offices

| | Total number of staff members | Proportion of personal staffs based on districts |
|---|---|---|
| 1972 | 1189 | 22.5 per cent |
| 1979 | 2445 | 34.6 per cent |

*Source:* John F. Bibby, Thomas E. Mann, and Norman Ornstein, *Vital Statistics on Congress 1980*, Washington, DC: American Enterprise Institute, 1980, p. 69.

and have more contact with them, than a George Miller of the 1940s could. For the parties this change in the political resources of congressmen has been important because it has made the constituency service of the individual public official more evident to the electorate; it has enabled the incumbent to claim personal support from voters on the basis of what he has done for them or their acquaintances. It can be argued moreover that the expansion of government services, first in the New Deal and then in the Great Society era, has increased government contact with citizens and has thereby increased greatly the number of potential clients for congressmen-as-troubleshooters.[18] To the extent that they adopted this service-orientation, not only were congressmen able to loosen their ties with the party ticket, they actually had an incentive to appear independent of it whenever it might be unpopular. Very rarely did the performances of these services depend on the co-operation of party organizations. Once again, political scientists have not collected the kind of data which enables us to say how much this change in the role of congressmen over thirty years has affected the propensity for ticket-splitting in the electorate. All we can observe is that the change in the role has occurred at the same time as ticket-splitting has increased and the marginal congressional district has become an endangered species.

As we have noted, congressmen were not prime examples of 'dirty garbage' even in Shorenstein's time. His remark was made to a state Assembly candidate, and it was candidates at this level of office, together with those for other lesser offices (of both state and local government), who were the principal examples of political flotsam. Indeed, it could be argued that state legislators were the most important element in the system. Party control over them at the county level, because of their lack of independent electoral resources, enabled county leaders to secure benefits for their locales through co-operation with the governor and legislative leaders. If this did not produce party government, there were in many states arrangements which were far from being instances of extreme decentralization of power. Just as congressmen have acquired the staff assistance, mailing rights, and free travel which have helped to make them more autonomous, so in many states have legislators acquired these facilities, and so too have they been able to use campaign techniques other than those supplied by their parties. To be sure, not all state legislators were in this position in the 1970s; in states such as New Hampshire the offices were still part-time ones with small remuneration. Even some of the larger states with full-time legislators did not provide their members with individual staff facilities; in 1982 Jewell pointed out that Massachussetts and Ohio had not done so.[19] But in many of the wealthier and more industrialized states the legislatures had become fully professionalized, and this made the job of legislator more attractive and provided the office-holder with independent political resources. Of the states in our study, two (California and New York) had legislatures which could be described as 'almost professional' thirty or forty years ago; by the end of the 1970s their members had resources which exceeded those available to congressmen in the 1940s. In Colorado, though, the state legislature remained, in theory, a part-time body, but here too there was change. As in many medium-sized states, legislators in Colorado were having to devote more time to their work, and their office facilities were improving.

Staff members, free journeys home to their districts, and free postage could not be used *directly*, of course, for campaign purposes, but indirectly they could be employed to the advantage of incumbents. For example, in New York state legislature members came to have the right to send three free communications a year to each constituent. Not surprisingly, one ex-legislator acknowledged that in election years most legislators chose to exercise this right just before primary or general elections, even though they could not include political material in their district newsletters.[20] Through increased name recognition among voters, they gained an advantage over their electoral opponents. Another benefit to the incumbent was the use he could make of some of his staff positions. In the case of congressmen and legislators in the largest states, at least one or two aides from his first election campaign could be rewarded with employment as staffers; they might want the job itself or, in some cases, they might value it as a stepping-stone in their own political careers. The promise of employment is a resource to be used by both incumbent and challenger alike, but the incumbent has the advantage of actually having provided employment to his aides. Those members of staff whom he wanted to work on his next election campaign could then take unpaid leave from their jobs, and could be reassigned to paid positions in the campaign organization. By the end of the 1970s, many legislators at federal and state level were able to retain aides who had campaign expertise so that they could deploy them at future elections.

As a crude summary, then, we could say that many legislators could now use public money to pay a kind of retaining fee to some of the purveyors of electoral services. It is a form of patronage politics which differed in two main respects from the more traditional form. First, even the most generous legislatures were not granting facilities which enabled the members to retain more than one or two campaign experts each. Secondly, with some notable exceptions, little of this patronage came to be controlled by party organizations; the individual office-holders clung to it jealously. In many respects Sandre Swanson, a member of congressman Ron Dellums's staff from 1973, was typical of the new breed of patronage appointee. He first became involved in political organization in the black community in 1970, the year Dellums defeated the congressional incumbent. He was then elected as a representative to a board set up under the community action programme, served as a volunteer in the 1972 Dellums campaign, and was asked subsequently to join the Dellums staff. The duties for which he was employed by Dellums involved casework for individual constituents and, as in many congressional offices, his employer seems to have been scrupulous in ensuring that political tasks were not performed in time allocated for office duties. Especially after Watergate, no congressman wanted to invite the scandal about the misuse of public funds that would surely follow if official and political activities were merged.[21] Nevertheless, like most district-based staffers, Swanson was extremely interested in politics; his spare-time activities were of direct use to his employer, and the contacts he made in conducting official business could benefit later his electoral mobilization activities. Consequently, Swanson was often removed from the paid staff at the approach of an election, so that he could work full-time as a political organizer. This started within eighteen months of his joining Dellums's office: in 1974 he was seconded to the Democrats' voter registration drive in Alameda County.

The provision of staff assistance to enable legislators to carry out district service functions for their constituents was not confined to Congress or even to legislatures in the largest states. Some city councils also started to vote for introducing this kind of facility for their members, and this aspect of 'professionalization' is not confined to the largest cities, such as New York. Of the cities in our study, Berkeley had also started to provide staff aid to individual council members. Of course, the less important the office, and the fewer aides each office-holder is permitted, the less likely it is that he can use these jobs to retain a campaign expert for his next election. The Berkeley city council members, with one aide each, would seek primarily someone who was competent in casework and office administration, rather than a low-level, would-be political consultant. At higher levels of office, however, even council members could make indirect electoral use of their aides because they could claim credit for the quality of the constituency service they provided. Nevertheless, it must be remembered that, as at the state level, the growth of the personal aide system varied from place to place; for example, the large cities of Denver and Oakland had still not introduced them by the end of the 1970s.

Not surprisingly, there is no evidence that, when aides are provided, legislators fail to make use of them to develop the casework side of their jobs. As Jewell has noted, 'where the legislator has staff assistance, he is more likely to give priority to constituency service, because this can be done without a great drain on his time'.[22] Not only does this service enhance his prospects of remaining in office whatever the fortunes of his party in other elections, it also enables him to free himself from a major part of the system of mutual obligations which tied his predecessors to their city party organizations. This is most clearly seen in the case of New York's state legislators. Indeed, the most striking aspect of the aide system in New York is the considerable impact it may well have had in advancing the breakdown of the traditional relationship between the legislators and political clubs.

One aspect of that relationship was reflected in the advice Hymie Shorenstein gave the worried New York Assembly candidate: campaigning outside of the campaign co-ordinated by the party bosses was insignificant. The other side of the relationship was that the legislators served the club leaders whenever the latter needed help in resolving problems faced by constituents. Those in need of assistance from the politically influential went to the party organizations, the clubs, for this, rather than approaching elected public officials directly. The clubs were keen to protect the gatekeeper role they played, and the legislators had no publicly-funded staff assistance to provide a constituency service of their own. In the last twenty years the decline of New York clubs has been paralleled by the growth of district servicing by the legislators. In a postal survey of all assemblymen and state senators who had districts in New York City every respondent reported that he kept at least one district office.[23] Indeed, about a quarter of the respondents (sixteen out of sixty-six) said that they had two or three district offices. But the most interesting result of the survey was the low level of involvement of political clubs in this aspect of the legislators' relations with their districts. Only three of the legislators had district offices on premises which were shared with a political club; and, of the twelve legislators who said they did share the premises of another organization, eight of them identified this as being a community organization, so that the legislator who shared with a club was unusual even within this group. Moreover, the clubs generated remarkably little business for the district offices. Asked 'Approximately what proportion of the constituents that are referred to your district office have first contacted an official or a member of a political club?', eleven legislators (20 per cent of the respondents) replied 'none', fifty-three replied 'less than 25 per cent', while only two of them (3 per cent of the respondents) claimed that 'between 25 per cent and 50 per cent' had done so.

The picture that emerges is of the typical New York legislator becoming relatively isolated in his district service work. As we have seen, less than one in five had premises which they shared with other organizations.[24] Moreover, a very similar number (fifty-four out of sixty-five respondents) reported that they did not share any of their district offices with other elected officials, such as a congressman, another state legislator, or a city councilman. Even in handling the problems bought by constituents, co-operation with other politicians has been the exceptional, rather than the usual, practice. Asked 'What proportion of the constituents' cases handled by your district staff involve close co-operation with the district staff of other elected public officials?', they replied as follows: five – 'none', forty-four – 'less than 25 per cent', fifteen – 'between 25 per cent and 50 per cent', and two – 'more than 50 per cent'. Of course, this isolation must be placed in context. It indicates very clearly the demise of the traditional role of party organizations; it provides some basis for the state legislator, like the congressman, to claim credit at election time for performing services, thus further weakening his links with the party. But it did not directly affect the party organizations' other functions. It did not reduce the legislators' dependence on the clubs for gathering election petition signatures, it did not weaken the patronage power of the Speaker of the Assembly, and it did not generate election workers for any legislator who fell foul of the dominant club in his district. Nevertheless, it is undeniable that it has been one of the factors which has led to a fragmentation of power in what was one of the more hierarchical parties in America forty years ago.

We have argued that in some instances, especially when two or more aides could be employed, public money has been used to maintain campaign experts in full-time employment. Although they have provided an advantage to incumbents *vis-à-vis* challengers, the aides employed in this way have generally remained outside public controversy. There are two reasons for this. Constituency liaison jobs have been widely perceived as valuable, because they have facilitated citizen access to government and this has overshadowed their contribution to the growing imbalance in political resources between incumbents and challengers. Another reason is that congressmen and state legislators have been able to reward only a few individuals in this way, and this has made it difficult for charges about the growth of personal 'machines' to be accepted. However, not all uses of staff allowances have remained uncontroversial. One controversial example is the part-time employment given to political clubhouse members on a revolving basis, a device used, for example, by several New York congressmen. The members were employed for a few months each to attend clubhouse functions, and to report to the congressmen on these meetings, for which they received between about $1,500 and $2,500; at the end of this period the job was handed on to another member of the club. Its practitioners defended the scheme on the usual ground on which district staffs were defended – it helped to keep the congressman 'in touch with his district'. Nevertheless, there were obvious differences between this and the sort of job performed by a Sandre Swanson: the work of the clubhouse members was tangential to the main casework of the district office, and the opinions with which they supposedly kept their employer in touch were solely those of party organization activists.

Brooklyn congressman Leo Zeferetti was one legislator who practised this form of patronage; in the mid-1970s he employed twenty-four people annually on a part-time basis, retained each of them for less than ninety days, and disbursed a total of $43,000 to them.[25] But like Representatives Scheuer and Biaggi who had also adopted the scheme, Zeferetti did not have his original political base in the clubhouses.

This point is important, because it helps to emphasize that, even when they could appropriate new forms of patronage, the party organizations were merely one of several actors in the electoral process. All three congressmen had risen to their positions independently of the regular party organizations, though none of them was an insurgent associated with the New Democratic Coalition. Without the reformers' 'clean government' ideology, yet with only insecure links to the party regulars, such congressmen naturally understood the need to repay political debts to the regulars in a manner that was most appreciated by them. In effect, what the congressmen were doing was using a resource which was normally employed to strengthen the incumbent's own political base as a means of tying the party organization to themselves. They needed the organizations' support because they could not expand their own organizations sufficiently to be absolutely sure of countering a potential threat from the clubs. Thus, while in most cases increased staff facilities helped to weaken ties to party, we see here an instance where they partially helped to preserve the party connection.[26] In the next section we see an even more unexpected development; in the somewhat unusual circumstances of California politics, the aide system has not helped to maintain an atomized political system, in which each candidate contests his election independently of fellow party candidates. Instead, it has led to widespread intervention by incumbents in the campaigns of others.

### 3. 'Warlordism' in California

For many years California was known as the home of weak parties, and of party candidates who largely controlled their own electoral destinies with little help either from party organizations or from other candidates. However, after the period of involvement by extra-legal party organizations in the 1950s, California's Democratic party did not return to a style of politics in which the individual candidate was an entirely autonomous, and dominant, actor in the electoral process. Rather, there developed a style which is best described as 'warlordism'. Three factors contributed to its development. One was the high cost of elections for the state Assembly with candidates facing competition from many major office candidates in the raising of funds. In each election year there have been some very expensive Assembly campaigns, and in these contests candidates have had to seek help in raising money and other resources. Thus, while the average amount spent by winning candidates in the contested general elections of 1978 was $43,000, twenty-three candidates that year spent over $100,000 each in their general election campaigns. A second factor was the Speakership of the Assembly which was powerful, and around which factions have developed. This provided the potential for intervention in electoral contests by a party hierarchy. This intervention developed after changes initiated by Jesse Unruh during his term as Speaker between 1961 and 1969. Not only did Unruh professionalize the legislature, he also introduced a system of financial contributions by interest groups and individuals which centred on the Speaker. In 1978 the Assembly leadership raised over half a million dollars which was distributed to Assembly candidates.[27]

What has emerged in California is not just intervention by the Speaker, nor merely transfers of money, but more wide-ranging electoral intervention. A third element of this new system has been the political expert-cum-aide. Not only have staffers often taken leave to run their employer's campaigns, they are also loaned to other candidates, especially those running for offices with constituencies within the legislator's own district. Congressmen, as well as state legislators, have been enmeshed in this network, and electoral intervention has extended to local elections. The loan of staff could take the form of the aide simply working in his spare time for the other candidate, or it could entail him being taken off the legislator's payroll so that he could work on the campaign full-time. The loaning of expertise, whether for general campaign management or for more specialized tasks, such as fund-raising, has created networks of obligation among public officials which bypass any party organizations. It is not surprising that these networks thrived in California where traditionally party organizations were weak and where the new campaign technology was first exploited.

The experience of Sandre Swanson exposes some important features of the new practice. In 1977 he took leave of absence from the Dellums staff to serve as the campaign manager for Lionel Wilson in the latter's successful bid to become mayor of Oakland. As a judge, Wilson had not run for any other public office in recent years, and thus did not have associates from previous campaigns to whom he could turn in staffing his campaign organization. In this somewhat exceptional circumstance for such a major office, an aide was lent to perform a general, rather than a specialized, role in the campaign. The Wilson case was also atypical in that virtually all the leading East Bay Democrats were enthusiastic about his recruitment, and most were willing to offer him campaign resources. Given that he was seen as having an excellent chance of winning, to fail to contribute resources was the more risky strategy in alliance building. Of course, after his victory Wilson's obligations to particular elected public officials proved difficult for them to redeem, because so many of them had contributed resources. There were two consequences of this for the local elections in 1979. On the one hand, as one East Bay political 'insider' put it, Wilson generally went 'endorsement crazy' that year in lending his name to the campaigns of others as a way of paying off his 'debts'.[28] On the other hand, in some crucial elections repaying one mentor meant alienating another, and in 1979 Swanson's employer, Ron Dellums, and his political allies found themselves unable to get the Wilson endorsement for their candidates.

As we have noted, the growth of the aide system at the state level dates from the early 1960s when Jesse Unruh initiated his policy of professionalizing the legislature.[29] Part of this reform involved the introduction of salaries appropriate to legislators who were doing a full-time job, but the aides were the most important element of the reforms. It is useful, however, to distinguish between the two kinds of expert who have entered California legislative politics since the 1960s, although both contributed to the growth of 'warlordism'.

The first, and more visible, experts were the aides based in Sacramento. Their number increased from about 115 in 1960 to over 700 by the late 1970s.[30] They have been employed to provide policy advice to legislators, but their contacts with the politically influential have given them the ideal base for building their own political careers; whereas in 1960 no state legislator had had previous experience as a legislative aide, by 1977 twenty-five serving legislators were ex-aides. In fact, in 1980 all three of the assemblymen from the northern East Bay were former aides. The Sacramento contacts gave the aides access to funds from the Speaker, other senior legislators, and state-wide interest groups. When they had local political contacts, and these were best developed when they worked for legislators from their home districts, the aides had another major resource for conducting campaigns of their own. This is not to say that aides have been disloyal and run against their previous employers, for normally this would be to court electoral disaster. The usual pattern has been for the aide to wait for his employer to retire, die, or run for another office (especially Congress), and then attempt to succeed him. Cultivating the district for himself has been as necessary for the ambitious aide as establishing himself with the political élite in Sacramento.

Of course, there are two main differences between these aides and congressional staffers based in Washington. First, the Sacramento political world is sufficiently small, and the relative size of the Speaker's political funds (and those of others) sufficiently large that contacts established there are electorally important. Washington provides few such advantages for staffers. Secondly, staffers have been able to travel back regularly to their districts on both legislative and political business for themselves, but members of Washington staffs have been able to do this rarely. Now the effect of the rise of the legislative aide in Sacramento, as a position on the political ladder in the state, has been to increase the networks which link the electoral activity of one public official to those of another. This link has been considerably strengthened by the emergence of the second kind of staffer – the district-based aide.

These aides are often part-time employees and, when they are full-time, California is not so generous as to provide its legislators with the same level of staff funds as that provided by Congress. Nevertheless, many state legislators have been able to retain one campaign expert on their staff to undertake the type of constituency liaison work which might involve two or three people in the case of a congressional district office. In the East Bay two of the legislators employed women who were both widely regarded as expert fund-raisers. This kind of aide has lacked the direct contact with Sacramento, which has been the crucial factor in the electoral success of the other kind of ex-staffers. However, their expertise has made them indispensable to their employers – both defensively and offensively. Defensively, the fund-raisers have helped incumbents to acquire sufficient funds to deter serious competitors over the two (or four) year electoral cycle and to carry out the minimum necessary electoral activities. Not all legislators have pursued an overt offensive strategy – intervening actively in other election campaigns – but those who chose to do so relied heavily on the expert aides. What has made the offensive strategy possible, and hence what has encouraged the growth of 'warlordism' in California, has been that campaign costs vary greatly from one election to the next. In open districts and in districts where incumbents were vulnerable, election costs rose sharply with increased competition; this meant candidates were often willing to accept aid from other legislators. Once the new incumbent was entrenched, the contributing 'warlords' would not need to repeat their assistance at subsequent elections, and could thus move on to other elections in search of future allies in the state Assembly. The cost of these occasional, highly competitive, elections could be staggering, and the figures cited earlier for general election expenditures understate the total sums involved. In 1978, for example, Elihu Harris spent more than $130,000 in winning the Democratic primary in the poor, black district of west Oakland. Harris, though, was not one of the legislators who received the help of another assemblyman's aide/fundraiser – only funds were directed his way.

Thus, the ability to contribute money or campaign experts to the highly competitive election contests enabled the incumbents of other seats to establish informal networks of obligation. At the centre of this 'warlord' system has been the Speaker of the state Assembly, the position around which the principal factions in the majority party have developed. Speakers have used their electoral resources offensively to consolidate their power within the legislative party and, not surprisingly, they have been best placed to extract funds from the Sacramento lobby. But to the extent that they had access to large potential donors, had safe districts, and had fund-raising experts on their staffs, other legislators could practice 'warlordism'. At the end of the 1970s the most frequently cited estimate was that there were less than ten state legislators (of both parties) who were consistently active in the campaigns of other candidates for the state legislature. However, a rather larger number than this were actively involved in local elections within their districts. The few who did operate outside their districts did not confine their activities to their own region of the state. Moreover, Democrats would also intervene at the behest of the Speaker, in exchange for future favours. One instance of this was described by an aide to a California Democratic 'warlord' from the East Bay, following a secondment to a district in another part of the state:

> I just was the campaign manager for [legislative candidate *A*] who just got elected in [*X*]. And that was simply a result of [legislator *B*], my boss, talking to the Speaker, and the Speaker calling me and saying 'We need to win; it's really important'. So that I left the district office for six weeks and worked there. I went on leave of absence, and not only raised the money – or helped to raise the money – but managed the campaign, bought the radio time, and wrote the brochures.[31]

This type of arrangement between *B* and the Speaker was essentially a short-term one. *B* had aligned himself with a group of Democratic legislators which was supporting the Speaker against his intra-party opponents, and his intervention in support of *A* strengthened his bargaining position *vis-à-vis* both *A* and the Speaker. Competition for control of the leadership precluded more permanent arrangements. However, such arrangements could develop in a vertical, rather than a horizontal, direction – that is, they could involve incumbents and groups at different levels of elected public office. One instance of this was the organization centred on congressman Henry Waxman in southern California, while another example was the co-operation established between Ron Dellums, Tom Bates, and the BCA organization, which began with Bates's election to the Assembly in 1976.[32] In the latter case permanent co-operation was facilitated by shared political ideologies which were the least attractive to potential large financial donors. This last point must be qualified in one respect – the Hollywood connection made some funds available. In particular, Jane Fonda used her popularity to attract contributors at fund-raising events for 'progressive' causes in the East Bay.

There was a further difference between Bates and his East Bay 'progressive' allies and most of the other practitioners of 'warlord' politics in California. For the most part, the others tried to loosen their ties to local activists and organizations to preserve the maximum possible freedom of manoeuvre in the making and unmaking of alliances in the state legislature. Dellums, Bates, and Alameda County Supervisor John George did the opposite – they set up an 'Executive Committee' each; these were groups of up to forty people whom they consulted on policy matters. Though none of them felt bound to accept any advice he received, the process of consultation itself impeded strategic manoeuvering on policies. The rationale for these committees was that they developed participation in the political process, a goal which all three officials supported, and that on occasions useful information and opinions could be forthcoming at Committee meetings. It should be noted, however, that they would have been useful for a rather different reason. In an area of California which was peculiarly well organized at the community level of politics, and with media campaigning rather underdeveloped there, public officials who were not sensitive to the views of their supporters might face more electoral difficulties than would candidates elsewhere. The Executive Committees can be seen as devices which served two purposes for the office-holders. They made it easier for sources of potential opposition to be identified early, and the very act of consultation helped to bind influential activists to the office-holder. That is, they would help to generate loyalty to the individual official; they were the functional equivalent of the revolving staff jobs used by the three New York congressmen mentioned earlier. If Executive Committees were an unusual device in California's 'warlord' politics, they did help to provide for the kind of political stability necessary for 'warlordism' to be practised at all.

In the last twenty years, then, there has been an important change in the role of elected officials in the Californian electoral process. What was formerly a system of 'each person for himself', with individual candidates working largely on their own, became a system in which some powerful incumbents would intervene regularly in other election contests. The extent of the change should not be exaggerated; intervention is possible in only a few contests each election year, special elections often provide the best opportunity for intervention, and local circumstances may preclude extensive intervention. Nevertheless, on balance, it must be admitted that the state has moved away from the kind of extreme fragmentation of partisan politics for which it was once famous. The California experience can be contrasted with that of New York. In New York fights between party factions have always been a major element of party politics, with factions putting up slates against each other at primary elections. In one sense, all that happened there in recent years were two alterations to existing practices. The factions became much looser, so that incumbents were now intervening more as individuals than members of a faction; and the intra-party conflict was much more likely to be pursued at the general election stage, with disloyalty to the party going unpunished. Because it had always been part of New York party politics, this intervention in other elections continued, while in California there were strict limits as to the kind of intervention which was possible. By the end of the 1970s, however, we do not find completely 'atomized' competition for public offices, in which incumbents and challengers remained aloof from competition for other offices, in either of these states. Yet, it is this kind of competition which we might expect to emerge as candidates become more independent of their party organizations, and, indeed, it has developed among the Democrats in Denver. At first glance, this transformation to fully 'atomized' competition might seem curious, since we might suppose that, in a state with such highly developed partisan conflict in both state and local politics before the 1960s, some forms of intervention might have survived. Three sets of factors seem to be relevant in explaining the absence of 'warlordism' and other forms of intervention. Future studies of other parties might consider the relative importance of each of them in explaining variations in intervention during the period of party decline.

The first factors are institutional. In Denver the seats in the state legislature were relatively unimportant as sources of power, and the legislators had small electorates which permitted the retention of older styles of campaigning. Colorado state legislators remained part-time public officials. Their conditions of service improved though, at $12,000 in 1979, the annual salary was still not that of a full-time public servant, even if (in real terms) this was double the legislative salary in 1970. As in most other state legislatures, there were improvements in office facilities at the state capitol: before the mid-1970s there were no facilities, but then a large room in the Capitol building was converted so that each legislator had a cubicle with a desk at which he could write letters and keep a filing cabinet. No provision was made for legislators to maintain their own district staffs. One or two of the Denver legislators did have district offices, but these were mostly situated in temporarily vacant shop premises. They could be staffed on a 'pool' basis between sessions by Assembly personnel, but otherwise the public purse provided no advantages for incumbents in 'selling' their constituency services to voters. Along with their part-time status, Denver legislators had few constituents to reach at elections; the typical Colorado assembly district had about 35,000 inhabitants, while the typical New York district had about two and a half times as many people, and the typical California district more than six times as many. In Denver extensive use of many of the newer campaign techniques was restricted, partly because they were too expensive for the kind of office involved, and partly because constituents still expected some form of personal contact with the state legislator, an expectation which necessitated older styles of campaigning. In these conditions the Denver legislator was not in a position to practise 'warlord' politics on a large scale. In comparison with assemblymen in the other two states, the Denver legislator was insignificant; in organizing a fund-raising event he was not in direct competition with the member of Congress, because the lack of importance of his position and the size of his district placed him in a lower league of fund-raisers; and fund-raising was less important for him than other campaign-related activities because he was expected to maintain more direct contact with people who were, in effect, fairly close neighbours. If the state legislators were not in a position to interfere in the elections of others, then it might be argued that 'warlordism' was much less likely to develop. Members of Congress are unlikely to be the initiators of such a style and often local government offices are not sufficiently important for incumbents to venture outside their own political backyards. In other words, state legislature politics are the key to the development of 'warlordism' of the California kind.

In these conditions, and with a Speakership that was much less important than the one in California as a fund-raiser, both conflict and co-operation between legislators in the electoral arena was limited in Denver. Legislators did not openly intervene in nomination disputes in other districts. At the general election stage they did not intervene against fellow Democrats, nor did challengers suffer from the endorsement of Republican opponents by leading Democrats – the kind of disloyalty found in New York. However, co-operation was equally confined. Safe incumbents did make small financial contributions to the campaign funds of fellow Democrats running in marginal seats, but this was more of a courtesy than a political investment of the California kind. Moreover, despite its history of party campaigning, Denver did not have joint campaigning by individual candidates; as in California, candidates preferred not to be linked to anyone else.

The second factor is the particular political circumstances facing the two key incumbents at the crucial stage of party organization decline. Intervention in electoral politics in Denver in the 1970s would have been dangerous for both the mayor and the member of Congress, and these were the only two offices from which a more interventionist style could have begun to develop. From the discussion in Chapter 5, it should be clear why Mayor McNichols refrained from any direct involvement in either state legislature or city council elections. After he had helped to frustrate Democratic party plans for partisan mayoral elections, McNichols's political survival depended on separating himself from other factional disputes in the party. To endorse one candidate over another in a primary election for the state Assembly, for example, would have been to put his own electoral strategy at risk. Rather different circumstances faced Denver's member of Congress, Patricia Schroeder. She regained the seat for the Democrats in 1972 from the one-term Republican incumbent. Schroeder was unlike McNichols; she had not been closely involved in party organization politics. Indeed, when she announced in 1972 that she would run against a long-serving state legislator for the congressional nomination, even a number of the liberal leaders in the party hierarchy regarded her as an outsider who was threatening the traditional role of the party in influencing nominations for public office. That the at-large districts had been abolished, that McNichols had frustrated party control over local elections, and that the power of the party had been successfully challenged in the revolt against Byron Rogers in 1970, did not undermine the view that Schroeder was an upstart. As we have noted, it took a number of years for those active in the party organization to realize that its power base had collapsed in the 1960s. One of the principal consequences of this was that Schroeder never developed the kind of close relationship with the party organization which might have led to her becoming involved in nomination controversies. Equally, after 1972 she was sufficiently secure from any challenges from within the party not to have to resort to entanglements in intra-party conflicts as a way of protecting her own position.

The third factor relates to the tradition of intervention in elections in Denver, which was different from both the East Bay and New York. In Denver the party organizations had been the central actors in the nomination and election process, but it was, in a sense, a highly impersonal involvement. Disputes had rarely become disputes between factions associated with particular individuals: individual candidates rarely endorsed each other; the party dropped state legislators from first-line designation in the primary only if they were extremely disloyal to the party, not because they represented a minority faction or group in the party; and the county chairman, while not a mere 'weighter' of different interests, was not permitted to destroy or systematically undermine any faction. In brief, party politics among the Denver Democrats was played according to certain rules which permitted extensive partisan activity, but minimized personal conflict within the party. The divisions within the party in the late 1960s and early 1970s were partly about the breaking of these rules: this was certainly the case with the resentment over the challenge to Rogers in 1970 and with Richard Young's mobilization of the party against McNichols in 1971. But that the rules might have been broken on these occasions did not alter immediately a political ethos that party politicians should not undermine their fellows. In the new era in which the party had relatively little power, candidates continued to respect the old rules.

These were not the rules used in either the East Bay or New York. In the former the very weakness of political organizations had long since forced politicians to make use of endorsements as a means of mobilizing activists; if it was one of the devices to be employed later by 'warlord' politicians, it had its origins in the incapacity of the parties. In New York endorsement and the creation of factional slates reflected a very different tradition: a tradition in which party loyalty was ruthlessly enforced and in which dissidents would be suppressed by virtually any means available. In New York clubs and individuals aligned themselves in factions, and only rarely did they seek independence from these larger groupings. One early instance of reformers going outside the 'slate' approach to campaigning was in 1965. Ed Koch, who was running against Carmine DeSapio for the district leadership of the party, announced that his name would appear in a separate column on the ballot from that of the reform mayoral candidate, William Ryan. He justified this on the ground that he needed more votes to defeat De Sapio than those which would be provided by Ryan supporters.[33] After the mid-1960s alliances in New York became much more loose than the slates of the earlier period, and endorsements by individual candidates replaced the cohesive slates.

We have argued that, despite its tradition of a relatively strong party organization, Democratic party politics in Denver was later characterized by a weak party role in the nomination and election processes and by non-interference in these processes by candidates for other public offices. Nevertheless, while this does provide an important contrast with California's 'warlord' politics, we must be careful not to exaggerate or misunderstand the differences in practice between the two areas. For example, although neither the county party organization nor the district party in Denver had much power to stop the self-recruited candidate who chose to direct his attention mainly at the primary electorate, the district captains and precinct committeepeople could still play a role in candidate recruitment. They were far more likely to play this role in safe Republican districts than in safe Democratic ones, but their involvement in recruiting candidates was not confined to 'unwinnable' districts. One former Democratic candidate said that she only agreed to run in her marginal district because of the persistence of the captains and the precinct committee members in their approaches to her.[34] This is the kind of intervention which was not possible in parties, such as California's, which had no base at the district level.

Even with the least costly activity for an 'interventionist' candidate, endorsing another candidate, most East Bay candidates did not look for a side to support in every election. Such a strategy would virtually have guaranteed the alienation of the maximum number of potential supporters. Obviously, aid was given most safely to a candidate who faced merely token opposition in the primary and who appeared to be in at least a competitive position for the general election. Seriously contested primaries brought much greater dangers. Consequently, some politicians who might have acted as 'warlords' did not do so; one of the senior East Bay legislators was renowned for not becoming entangled in controversial elections involving two or more Democrats. Furthermore, one of his 'warlord' colleagues did not publicly intervene in the 1979 three-candidate primary in west Oakland until it was apparent that Elihu Harris had a clear advantage in organization and funding; he then endorsed Harris who went on to win the Assembly seat.[35] Even in local elections many legislators preferred not to be drawn into an endorsement until they were sure that they were backing a winner whose support in the future might be useful to them. In 1979, when Oakland Democrats were split badly over a contest for a city council seat, some major public officials, including one of the local 'warlords', did not give any endorsement until the run-off election stage, by which time the strength of the two campaigns was known. Indeed, some incumbents in the days of the cross-filing primaries, when endorsements were important in mobilizing activist support, had always chosen to remain aloof from other elections; one ex-legislator reported that he did not make a habit of endorsing other candidates, except in the case of the head of the party ticket.[36] In the 1970s most politicians still subscribed to the view that guided many of their predecessors – endorsing someone meant you would share more of his political enemies than his friends. For those who were electorally secure or politically less ambitious this view led to a policy of limited endorsement. For others, including the 'warlords', it was still a consideration to be borne in mind in particular cases but it did not guide their actions.

To conclude our discussion of electoral intervention in California we cite two instances to illustrate the complexity of endorsements, and hence the difficulties confronting 'warlords' contemplating interventions. One legislative candidate, who had fought a primary as an 'underdog', said that despite his position in the contest he did not want the endorsement of a local congressman. He could have received it, but he asked that it not be given because of the latter's controversial stance on a major public issue of the time.[37] Successful intervention requires that a combatant will find the endorsement, or other assistance, useful. The author was also present at a campaign meeting when a decision was made about a possible endorsement by a candidate for non-partisan elective office in the East Bay. The candidate, *A*, and his closest supporters discussed whether to enter into a cross-endorsement arrangement with *B* who was running for another non-partisan office in the area. *B*'s opponent, *C*, had already endorsed *A*'s principal opponent, so that the question of *A* endorsing *C* did not arise. *A* argued that many of *B*'s potential supporters would vote for *A* irrespective of whether they endorsed each other. On the other hand, while some of *C*'s supporters might vote for *A* if he did not endorse *B*, they were less likely to do so if he made the endorsement. This argument produced agreement among all his supporters that he should not endorse *B*. (In fact, *A* and *C* won the two elections.) Although they are the least costly form of electoral assistance, endorsements still entail risk-taking, so that intervention in other elections necessitates an understanding of local conditions. This is one of the reasons why 'vertical' intervention by state legislators is that much more common than 'horizontal' intervention; they have greater knowledge of the conditions in their own districts.

In this chapter we have tried to show that the declining 'pull' of heads of party tickets, and the increasing resources available to candidates, has not produced a uniform pattern of 'atomized' electoral competition, centred on the individual office. As we might expect, there are instances of 'atomized' competition developing, but electoral politics of the Denver type was not the only form found at the beginning of the 1980s. On the one hand, we found in New York City that party decay has not proceeded so fast as to remove intra-party conflict from the electoral arena. Candidates in New York interfere in the elections of others just as party clubs do. On the other hand, in California a number of conditions combined to make a limited kind of electoral interference attractive to some incumbents; the main factor in promoting 'horizontal' interference has been the high cost of some state legislature elections. Paradoxically, the availability of the new campaign technologies has provided the opportunity for candidate independence, but it has also created the need for co-operation between public officials in some circumstances. In the next chapter we examine how candidates for Congress and for lesser offices have been able to make use of these technologies.

## Notes

1. Tolchin and Tolchin, *To the Victor*, 18–19.
2. The analysis of the 'system of 1896' was developed even further in the work

of Walter Dean Burnham; see especially *Critical Elections and the Mainsprings of American Politics* (New York: Norton, 1970) and *The Current Crisis in American Politics* (New York: Oxford University Press, 1982).

[3] Donald E. Stokes, 'Parties and the Nationalization of Electoral Forces', in William Nisbet Chambers and Walter Dean Burnham (eds.), *The American Party Systems* (New York: Oxford University Press, 1967), 201–2.

[4] Interview no. 96.

[5] John F. Bibby, Thomas E. Mann and Norman J. Ornstein, *Vital Statistics on Congress, 1980* (Washington, DC: American Enterprise Institute, 1980), p. 19.

[6] See, for example, Norman H. Nie, Sidney Verba and John R. Petrocik, *The Changing American Voter* (Cambridge, Mass.: Harvard University Press, 1976), pp. 52–3.

[7] One of the few sources of information on state legislate elections is Malcolm E. Jewell, *Representation in State Legislatures* (Lexington: University of Kentucky Press, 1982), ch. 2.

[8] One of the curious features of American political science in the last thirty years is that the relationship between party and candidate campaign organizations has not been the subject for much study. Although the reaction to the APSA Report did lead to considerable attention being given to the weakness of the parties in Congress, there was never much interest in the party link in electioneering.

[9] David R. Mayhew, *Congress: The Electoral Connection* (New Haven and London: Yale University Press, 1974), p. 33.

[10] The data for all the elections up to 1978 were taken from Bibby, *et al.*, *Vital Statistics on Congress, 1980*.

[11] The figures cited exclude incumbents who first won in a special election after the 1978 general election. They also exclude an incumbent (Myers of Pennsylvania) who won 72 per cent of the vote in 1978 and who was defeated by an independent candidate in the 1980 general election. In an earlier book we argued that, for Senate seats, having a large plurality (defined as receiving more than 55 per cent of the vote) did not increase an incumbent's prospects for winning the subsequent election; Ware, *The Logic of Party Democracy*, 59–61.

[12] Gary C. Jacobson and Samuel Kernell, *Strategy and Choice in Congressional Elections*, 2nd edn. (New Haven and London: Yale University Press, 1983).

[13] Jewell, *Representation in State Legislatures*, p. 45.

[14] Of course, Stokes is correct in suggesting that competing for attention was easier in small-town America of yesteryear, with the dissemination of news controlled by local editors, than it is in an era when national news dominates the broadcast media ('Parties and the Nationalization of Electoral Forces', p. 197.) Nevertheless, this was an advantage to the candidate only when his party organization's relations with both the editors and the electorate were weak.

[15] Harrison W. Fox, Jr., and Susan Webb Hammond, *Congressional Staffs* (New York: Free Press, 1977), p. 19.

[16] Interview nos. 43, 44, and 45.

[17] Richard F. Fenno, *Home Style* (Boston: Little, Brown, 1978).

[18] Morris P. Fiorina, 'The Case of the Vanishing Marginals: The Bureaucracy Did it', *American Political Science Review*, 71 (1977), 177–81.

[19] Jewell, *Representation in State Legislatures*, p. 143.

[20] Interview no. 133.

[21] Curiously, it is not illegal under federal law for an aide to work on a campaign while he is still on the federal payroll; *New York Times*, 10 Feb. 1981.

[22] Jewell, *Representation in State Legislatures*, p. 143.

[23] For details of the survey, see 'Bibliography and other Sources'.

[24] The question asked was: 'Do you share the use of any of the premises of your district offices with any of the following organizations – a political club, a labour union, a community organization, a religious organization, some other organization?'

[25] *New York Times*, 26 June 1976.

[26] Other instances, already mentioned in Chapter 4, are the $3,750 a year, 'no show' jobs which Brooklyn Democrats in the state legislature were supposed to create from their staff allowances.

[27] Jewell, *Representation in State Legislatures*, pp. 41–2.

[28] Interview no. 10.

[29] John Adkisson, 'Staff Assistant Today, Assembly Member Tomorrow', in T. Anthony Quinn and Ed Salzman (eds.), *California Public Administration* (Sacramento: California Journal Press, 1978), pp. 32–3.

[30] The information in this paragraph is derived partly from Adkisson, 'Staff Assistant Today, Assembly Member Tomorrow'.

[31] Interview no. 80.

[32] Bates's predecessor, and former employer, Ken Meade was much less interested in 'warlord' politics than Bates. His distance from both the Dellums organization and the BCA reflected disinterest in this aspect of politics, rather than an ideological incompatibility between him and his allies.

[33] *New York Times*, 17 Aug. 1965.

[34] Interview no. 95.

[35] Interview nos. 36 and 48.

[36] Interview no. 16.

[37] Interview no. 26.

# 7
# The Impact of the New Campaign Technology

Of all the aspects of decline in America's parties, the one most apparent to the layman has been the growth of new styles and techniques of campaigning by the individual candidates. He experiences directly the use made of television in campaigning, he may well be one of the millions of people receiving mail from candidates, either advertising themselves or soliciting financial contributions from him, and he might know the names of some of the campaign specialists and polling firms employed by the candidates.[1] As a result of this revolution in campaigning, the cost of elections increased dramatically after 1964; in 1952 total campaign spending in that presidential election year was \$140m., by 1964 it was \$200m., but by 1976 it had reached \$540m.[2] In fact, the increase of \$125m. between 1972 and 1976 would probably have been much greater, if the 1974 Amendments to the Federal Election Campaign Acts had not restricted the growth in expenditures on presidential elections through a number of devices. No one would doubt the validity of the layman's view about the changes in campaigning for the most visible offices – the presidency, the US Senate and the state governorships – but what is less obvious is the impact on other offices. These offices are either contested in districts smaller than a state, or are lesser offices contested on a state-wide basis. It would seem unlikely that television would have transformed elections to, say, city councils or even to many state legislatures, especially the less professional ones and those in the smaller states. On the other hand, it is clear that there has been a major change in campaigning for many, though not all, congressional seats. This is reflected in the increased expenditure on congressional elections in the 1970s – an increase of 34 per cent in real terms between 1972 and 1978.[3] By 1978 total expenditure in the average congressional district was more than \$200,000, but the variation between districts was great: it ranged from a few thousand dollars, where a safe incumbent was running, to \$1.7m.

In this chapter we examine the impact of the new campaign technologies on offices contested in areas smaller than a state – that is, offices such as congressional seats, state legislature seats, and posts in local government. Before doing so, it is worth remembering that, at all levels of office, one of the main effects of higher election expenditures was to diminish the role of party. Although there are no reliable statistics for the entire period 1960–80, to show either the proportion of total campaign expenditures made by the parties or the proportion of candidates' campaign income donated by parties, there can be no doubt that both declined dramatically. In the case of federal elections, legal changes in the 1970s limited the contribution a party could make to a campaign, and hence reduced the incentive for a party to emphasize fund-raising. In addition to the well-known revolution in presidential campaign funding, the new laws had a significant impact on the funding of congressional elections. Apart from stimulating the growth of political action committees, the reforms limited to $15,000 the contributions national and state party organizations could make to a candidate's campaign. With more expensive elections, a decline ensued in the share of the candidates' funds supplied by party organizations; in 1972 parties had supplied 17 per cent of House candidates' funds, but six years later they supplied only 8 per cent. (There was a similiar decline in the case of Senatorial candidates.)[4] Thus, even when they did not lose their fund-raising capacities for other reasons, county and state party organizations declined *relatively*, both because the stakes of campaign finance had been raised and their own contributions to federal elections had been restricted.

However, there were other pressures on the parties as suppliers of campaign funds. Activists were not very keen on fund-raising: those surveyed in Denver were even less enthusiastic about fund-raising for the party than they were for a candidate.[5] Moreover, unless there were special circumstances, as with some labour unions in Brooklyn and the Bronx, the decline of party ties meant that individuals and groups increasingly seem to have preferred giving money to candidates, rather than parties. Consequently, even the state party organizations, which were becoming more professionally organized in the 1970s, were having to survive on smaller budgets. For example, Huckshorn and Bibby have argued that the average budget for state parties, in their fifty-three state party sample, increased from $188,125 in 1960 to $340,600 in 1980; but, in real terms, this meant that in 1979 the average party had a budget which was only 73 per cent of the size of its 1960 budget.[6] This decline may be contrasted with the increase in campaign expenditures in America which, in real terms, rose by about 60 per cent between 1960 and 1976. Almost certainly, at the city level the reduced ability of parties to finance elections was greater still. In places such as Denver, where the party lost its power to coerce candidates into making contributions to its own campaign, the decline was marked. In the 1950s about one half of the money spent by state legislature candidates was given to the county party for its campaign. By 1978 the party received nothing from the candidates and had little to give them; even counting state party contributions, which were far greater, Democratic party contributions to the Denver candidates that year met less than 5 per cent of the candidates' expenses.

It is useful to begin our study of the technologies by examining, at a very general level, the considerations which influence a candidate for a 'local' office when he is deciding how important each of the available techniques should be in his own campaign. One consideration is the strength of the local party organization; mass media campaigning will be more important in areas where the organizations are weak. Wattenberg's study of congressional elections found that, where the organization was perceived to be strong by the campaign manager, the average campaign spent 54 per cent of its budget on mass media; but where the organization was perceived as weak, the average campaign committed 73 per cent of its budget to mass media.[7]

A second factor is tradition. Expectations on the part of voters as to how a campaign is properly conducted can make a radical departure a risky strategy. For example, it was widely held among Denver's political élite that some form of personal contact with voters, by the candidate or his campaign workers, was essential for a campaign to be acceptable to the public. In discussing the difference between Denver and two other areas in which he had been employed, Kansas City and Texas, one campaign consultant argued that door-to-door campaigning continued to be used in Denver because of the size of the electoral districts and tradition.[8] Not only does the past influence what resources have to be acquired, but it also affects how they are best used. In the 1952 congressional election, Colorado's 'moralistic' political culture disadvantaged a candidate whose techniques were acceptable in other American political traditions. That year Denver Democrat Byron Rogers was making his first bid for re-election, and he might have been vulnerable in a year when Eisenhower was the Republican presidential nominee. However, his Republican opponent, who was originally from Tennessee, employed two techniques which were largely unknown in Denver elections. He did virtually anything to gain public attention – he was the first congressional candidate to use television commercials in the city – and he criticized continually Rogers's integrity and judgement. Hostile press and public reactions to this behaviour enabled Rogers to win. Of course, traditions do change; familiarity with advertising by presidential candidates on television eliminated the hostility to the device when it was used in later congressional elections in Denver. The past still has enough influence on the present, however, to make campaigning in Denver differ from that in, say, Knoxville.[9]

A third factor is the level of office being sought. Although some wealthy candidates occasionally spend far more on a campaign for a particular type of office than anyone else has previously, there are reasonably well defined upper limits of campaign expenditure for all offices beyond which it is simply not rational to go. For that reason, the type of election campaign that can be planned for a city council seat in town *A* will more closely resemble other contests for council seats in *A*, than it will a congressional election or a US Senate election. We do not see extensive television campaigns for most lesser public offices, because not even a highly-financed, or an individually wealthy, candidate would choose to campaign in this way. Our point is not that such a campaign would be counter-productive, though charges of 'buying' the election might make it so, but that it is a waste of money for these offices when there are much cheaper techniques which are appropriate for them. For example, a Republican candidate for the Oakland city council in 1979 spent about $70,000 in a campaign that was considered a highly-financed one for an East Bay council seat; he did not advertise on television at all because the campaign could not afford it.[10]

A fourth factor affecting the mix of techniques chosen by a candidate is the relationship between his district's boundaries and the boundaries of the different media in which he might advertise. In this respect the main differences between television and newspapers are: (i) that the very high cost of television advertising means that any decision about its use is likely to be a central issue in planning a campaign; and (ii) there are no neighbourhood, and sometimes not even any city-based, stations which are equivalent to local newspapers. Obviously, television is most cost-effective for an election such as that for the Presidency, when all registered voters within the viewing area of the station may participate – the only 'wastage' for the candidate involves those who are ineligible to vote and those who have not registered to do so.[11] In many gubernatorial and US Senate elections the 'wastage' is also fairly low: in most states the television stations have viewers who are residents of the state. Nevertheless, there are some notable exceptions, such as New Jersey. In the northern part of New Jersey the viewers watch New York stations, while in the southern part of the state they receive transmissions from Philadelphia stations; there are no major commercial stations based in New Jersey itself. The position facing candidates in congressional elections is far more varied than it is for Senate candidates. At one extreme is Alaska where the single congressional election can scarcely be described as a 'local' election: the only television stations to which the inhabitants have access are ones broadcasting within the state. An Alaskan congressional candidate has no 'wastage' from television advertising; his advertisements are not seen by residents of other congressional districts. At the other extreme is New York City. Here in the 1970s the stations broadcast not only to seventeen districts in the city itself, but also to congressional districts in northern New Jersey and western Connecticut, as well as to those in eastern Long Island and to suburban counties to the north of the city. If a congressional candidate in a New York City district used television advertising, he did so knowing that only a small proportion of the viewing audience could vote for him. Naturally, in large urban areas, television has been even less suitable for campaigning in state legislature elections than it has been in congressional elections, because there are more districts being contested. Consequently, state legislature campaigns in which television has been used extensively have been found normally only in states with professional legislatures, and even there they have often been confined to districts outside the major cities. Television campaigning by state legislature candidates cannot account for the demise of the urban parties.

This leads us to a fifth factor: the type of district being contested. Proportionately, there are more new voters at each election in the newer suburbs than in other kinds of district, and in that environment previous party loyalties become less fixed. The suburban candidate would require more, not less, activist support to be as effective as his centre-city counterpart, whereas in fact voter turnover means that it is usually more difficult for him to generate activism. The result is that suburban candidates have to make more use of television, and other non-labour-intensive techniques, and this has been reflected in the election expenditures of these candidates. While the average congressional campaign in 1978 cost $155,000 in urban districts, and $176,000 in rural districts, it cost $197,000 in suburban districts.[12]

A sixth factor is that, even in areas where party workers are normally available, a particular candidate may lack them and, therefore, may have to turn to media campaigning to overcome his disadvantage. A good example here is the case of Edward Koch in the 1977 mayoral primary in New York City. Koch was faced by six opponents, all but one of whom had some form of mass organizational base. The access to campaign workers that each of their bases provided allowed the candidates some flexibility in choosing between different mixes of campaign techniques. Bella Abzug, for example, decided to use her workers in conjunction with an extensive 'telephone-bank' operation, so that the workers made personal calls on those who had already been contacted by phone.[13] Koch had very few campaign workers outside Manhattan, however, and even there he had divorced himself from much of his original East Side reform club base, by presenting himself as a 'moderate' candidate who was in favour of the death penalty. He could devote his campaign funds only to techniques which did not require volunteer workers – such as television, radio, and direct mail. On the advice of consultant David Garth, he chose to spend almost all the money – $600,000 before the run-off primary election – on television advertisements. Koch's two opponents from the minority communities faced the opposite problem: they had campaign workers but could not raise enough funds to run a television campaign. These two cases indicate an obvious, and important, seventh factor in determining the priority a candidate will give to different techniques. If fund-raising presents a problem for him, then he cannot make much use of the most expensive techniques, and in particular he is restricted in his ability to take advantage of television advertising.

What the preceding discussion suggests is that we cannot expect to find 'typical' campaigns for the US House, state legislatures, or city and county offices. The circumstances facing candidates vary between different regions of the country, and also within the same state, and the circumstances facing different candidates for the same office also vary enormously. This presents an important contrast with competition for the Presidency and most governorships and US Senate seats where campaigns have become remarkably similar. Our aim in this chapter is not so much to describe the variations which developed in the use of the major techniques, but rather to outline some of the incentives, and constraints, involved in the deployment of particular techniques. In doing so, not only can we explain the changes in the period up to 1980, but we can also explain the potential for further changes in styles of campaigning for lesser offices. We examine in turn the employment of campaign consultants, television, radio, the distribution of campaign literature, and door-to-door campaigning.

### 1. Campaign Consultants

One plausible, but false, hypothesis about the growth of campaign firms is that, since the experts utilize techniques which are expensive, they cannot be taken on in campaigns for lesser offices. Certainly, there are large numbers of electoral districts, especially for school boards and for seats on city councils, where consultants have never been employed. Yet, for two reasons, professional expertise has not been confined to expensive campaigns for major offices. First, even up to the level of a state legislator, a growing number of candidates have been political novices who did not know how to run an election campaign. Until the 1960s there were many states where members of the party organizations would take care of the novice candidates and help them to run their campaigns. It was only in states with highly disintegrated party structures that this kind of shepherding could not occur. However, the declining recruitment to party organizations, the inexperience of many precinct committee members who remained, and the growth of an ethos of working only for those candidates of whom one approved meant that often party organizations could not meet the demand for candidate training. Secondly, not all the services a consultant could supply have been expensive. One consultant in Denver cited computer analysis of voter turnout, by precinct, as an example of an inexpensive service; in the previous state legislature election the purchase of the relevant tapes and simple computer analysis had cost one of his candidates $170. For this sum the Democrat employing him had received information which had enabled him to target his campaign literature to those two-party competitive districts in which there was normally high voter turnout.[14]

Relatively elementary computer analysis can be purchased from many other sources, and most of the larger consultancy firms do not bother to take on such work, or do so only as a personal favour to a particular candidate. Nevertheless, there are services which can earn money for a firm even when the client is running for a lesser office. Opinion polls are the best example of this, perhaps. Although by 1980 many candidates for lesser offices still did not employ pollsters, the latter could provide an inexpensive service. One successful challenger for a city council seat in Denver commissioned a 700-person survey in his district before announcing his candidacy, and then bought two 100-person polls during the course of the campaign. All were telephone polls, and they cost him less than 5 per cent of his $10,000 campaign budget.[15] An even less costly technique is the adding-on of one or two questions to an opinion poll survey when the main client is, say, a congressional candidate; this is a practice known as 'piggy-backing'.[16] Obviously, the larger the size of the lesser candidate's district, in relation to that of the commissioner of the poll, the more detailed is the analysis which can be undertaken subsequently. The main point to be emphasized is that there are a number of services which many lesser office candidates could buy from campaign consultants, and which would not involve the consultants in making a loss. However, even by the end of the 1970s, this was an area which was far from being fully exploited.

That many candidates did not know how to organize a campaign in the late 1970s is scarcely surprising. The weakening of party organizations in the cities, and their failure to develop in most suburbs, greatly increased the likelihood that those running for positions in local government or for the state legislature would have had little campaign experience. Naturally, the increased involvement of political novices was most evident in seats which seemed unwinnable at a particular election. Experienced party workers were less likely to 'try their hand' as a candidate in these conditions, just as the politically ambitious were not attracted to such contests. Thus district captains and others often had to encourage relatively unpromising novices to contest a seat. It is important to realize that sometimes the seat in question was not an entirely safe seat for the opposition, but was merely seen as unwinnable in the context of a particular election. In fact, the most extreme example of an unskilled novice whom we interviewed during the course of the research had contested a district where four years earlier his party's candidate had secured 49 per cent of the vote. He was not a self-recruited candidate but one who had been persuaded to stand by local party officials.[17] It was this kind of candidate who was a potential purchaser of the 'how to run a campaign' packages marketed by some political consultants.

One firm which provided this kind of service at the end of the 1970s was Campaigns West, a firm working primarily within Colorado. A group of two, and sometimes three, individuals, all of whom had run for public office themselves, Campaigns West was an attempt to make a business out of what had been a past-time for its owners. Since a reputation for success is what generates most business in the consultancy industry, it is difficult for new firms to obtain clients. Nevertheless, Campaigns West was asked in 1978 to take on as clients a congressional candidate and an incumbent running for re-election to state-wide office. To expand its activity and hence the possibility of establishing a reputation, it also provided some services for candidates running for the state legislature however. One of the packages, aimed primarily at political novices, cost $600 and provided them with training in organizing campaigns, handling political issues, and targetting campaign literature. What was, perhaps, most interesting about this package was that in some cases Campaigns West actually supplied campaign managers to candidates. The managers were well-known Democratic party campaign workers who wanted to extend their political experience by running a campaign, and who were prepared to undertake the work for a nominal sum, $100–150. In effect, the firm was taking on one of the traditional party functions – linking the party candidate with an experienced political organizer – but a function which was now performed poorly by the Colorado parties themselves.[18] If it can be argued that the performance of this function is important for democratic politics, then it must also be admitted that it is not financially rewarding. Even four or five clients buying this package from Campaigns West were not going to make the firm profitable. It was really a device which served two related purposes: it made the firm well-known among the political élite in the Democratic party, and thereby built up goodwill, and it also helped to create an image of a firm engaged in a number of campaigns – that is, of a firm which could get business. (The use of the 'loss-leader' approach to campaign consultancy was taken even further in a case reported in southern California. There consultant Wayne Adelstein took over the campaign of a Republican primary candidate who had made no speeches and had undertaken no fund-raising or door-to-door campaigning. Adelstein acquired $40,000 for her in campaign contributions, which was spent mainly on distributing literature, and the candidate eventually finished second in the primary with over 30 per cent of the vote. Adelstein admitted that his intervention in the election was a gimmick to gain publicity for his firm.)[19] Ultimately, though, Campaigns West's survival depended on success with its major candidates; while both of them won that year, they gained only narrow victories, and Colorado's political élite did not view the firm as having been very effective. It went out of business the next year.

Because the Democrats in Denver usually had smaller campaign budgets for Assembly elections, and had more safe districts in which the Republicans fielded only token opposition, fewer Democrats than Republicans used campaign firms. Moreover, those Republicans who employed professional expertise normally bought more services than did the Democrats.[20] As with the Democrats, there was one Denver-based firm which handled campaigns for Republicans at all levels of public office from the US Congress downwards. Unlike Campaigns West, this firm's involvement in state legislature campaigns was more likely to consist of providing all the services for a campaign, rather than just a 'starting up' package followed by advice whenever it was requested. Most Republican candidates for the state Assembly wanted something less than a 'complete' package, however, and they were more likely to consult a 'generalist' public relations firm, employ a specialist polling organization, and engage an independent professional campaign manager, rather than use this campaign firm.[21] This point is important when considering the future of campaign consultants in campaigns for lesser offices. Most of the services needed by these candidates have been obtained either from large campaign firms, as with 'piggy-back' polls, or from firms for whom political work has been only a sideline, such as public relations firms which could design campaign literature. Unless lesser office candidates find it much easier to raise campaign finance, or there is a great increase in the value of election to offices such as the Colorado General Assembly, it seems unlikely that many candidates for these offices would be able to pay for a 'complete' package of campaign services. If the market at the state legislature level is small, then there is probably insufficient business in small and medium-sized states for firms to put much effort into expanding the lesser office sector. In other words, firms such as Campaigns West would be able to survive only if they expanded on a permanent basis into the major office market. This is not to suggest that the *slow* growth in the use of professional expertise will not continue, we are merely arguing that there would not seem to be a vast, untapped, market of small clients waiting for the right kind of entrepreneur to enter the market. It is unlikely, then, that campaign style and organization for lesser offices in the Colorados of America will change as rapidly, for example, as they did in gubernatorial elections between the 1950s and the 1970s.

Nevertheless, the slow change in the use of experts in campaigns for semi-professional state legislatures should not be ignored. It is to be expected that, increasingly, candidates will purchase opinion poll and voter analysis data from specialist firms. Furthermore, on occasions, generalist firms can be expected to provide other services for candidates in efforts to develop their reputations.[22] But perhaps the most interesting portent for the future was the increased use of paid campaign staff in Colorado in the 1970s. If Colorado's state legislature did not warrant the widespread employment of consultancy firms, there was a growth in the employment of what might be termed 'semi-professional' campaign managers, particularly in state senate elections.[23] One consultant suggested that the principal reason for this was the shrinking pool of available volunteers, especially women who 'have been automatic volunteers', and this meant that paid staff were a necessity in some campaigns.[24] They were a new kind of political actor: they did not earn their living by managing campaigns, but equally they did not have *quite* the same loyalties as either the amateur or the professional activist. They brought with them into the campaigns their 'baggage' of political beliefs, but their commitment tended to be to their candidate winning, rather than to a cause or to the party. Most of these 'semi-professionals' were experienced party activists who found that there was little scope for their abilities and enthusiasm in working solely for the party organization. Yet, unlike the campaign specialists in California, they could not become 'full professionals' employed by a state legislator, because Colorado made no provision for aides for its legislators. Nor, for the most part, did they have the range of skills or the motivation necessary to enter the consultancy business. Yet they were able to fill a major gap in electioneering created by the demise of the party organization. Thus, it is plausible to argue that in the future the 'semi-professional' campaign managers will become a more distinctive element in states, such as Colorado, with semi-professional legislatures: continuing to practise their craft, they will become more clearly separated from the party and issue activists.

That future changes in the use of experts are likely to be slow for many lesser offices was also evident in the experience of the East Bay. There too candidates for city councils, county boards of supervisors, and various elected boards of control purchased certain campaign services, but the employment of consultants providing a wide range of services was rare. Of course, for some candidates the availability of expertise through the 'warlord' system obviated the need for paid independent consultants, but this was not the main reason for their non-appearance. The fact was that some expert advice could be purchased in campaigns with budgets of between $5,000 and $30,000, but a fully professional campaign was impossible if enough money was to be left for literature, billboards and the rest. Yet the growth of the consultancy business in northern California was not confined to the congressional and state legislature sectors. There were other sources of business which were peculiarly well developed. One of the sources of new clients was special interest groups (including landlords and tobacco companies) who found themselves threatened by 'public interest' initiative referendums at the local level, and who turned to the consultants to run campaigns against the proposed laws. Funding was not a problem for these interests, but their inexperience in political campaigning meant that often they were content to buy a complete package from one consultancy firm. The other important area of expansion was local government where, until the mid-1970s, San Francisco members of the Board of Supervisors (the city-county council) were elected in a city-wide election. There were several crucial differences between San Francisco and Oakland which help to explain why the latter developed professional campaigning only in mayoral elections, while the former was one of the few cities in the country where this kind of campaigning extended to Supervisorial elections. First, San Francisco had about twice as many inhabitants (700,00) as Oakland. Secondly, like Denver, San Francisco was both a city and a county, so that there was only one set of local government politicians fighting for publicity. Thirdly, the principal Bay area newspapers, and all the television stations, were based in San Francisco and, while they gave little attention to politics in the East Bay, there was considerable coverage of San Francisco politics. These three considerations made possible fund-raising on a much larger scale than in the East Bay, and in turn this permitted television campaigning and the extensive use of consultants. It was not surprising, therefore, that there were a number of consultancy firms based in San Francisco, but none located in the East Bay. However, the circumstances which generated professional local government campaigns in San Francisco were unusual, and, even in states with professional legislatures, it seems unlikely that there will be much growth in fully professional campaigns at the level of city councils.

Finally, when we turn to congressional elections, it is important to remember that the high average cost of campaigns in the late 1970s concealed great variations between districts. Not all the congressional elections in any given year would produce campaigns of a scale which differentiated them from, say, campaigns for a semi-professional state legislature. In particular, when an incumbent appeared to have an impregnable position, both he and his opponent might run low-budget campaigns. The likelihood of this increased the fewer the sources of campaign finance available within the district. A typical example of the safe, poor-district, incumbent was Shirley Chisholm who, although opposed by Republicans, spent $14,000 and $31,500 respectively in winning the 1978 and 1980 elections in her Brooklyn district. Of course, low-funded campaigns for higher offices were not unknown; for example, Senator Proxmire spent a mere $697 in winning 72 per cent of the vote in Wisconsin in 1976, but these instances were less common because there are fewer safe Senate seats and governorships, and no safe Presidency. In low-budget campaigns when both candidates were, in effect, admitting that there was only a remote possibility that the challenger would win, there was little point in making much use of consultancy firms. Now it might be thought that, given the large number of safe congressional incumbents, there would be a corresponding number of low-budget campaigns. This was not so. In 1980 there were a mere forty-eight congressional districts in which total campaign expenditures at the general election were less than $50,000.[25] The reason for this was a natural caution among incumbents about their re-election, combined in some cases with attempts to ward off future challengers by displaying the resources available to an incumbent; in 1980, for example, Arkansas congressman Beryl Anthony spent over $100,000 in his campaign, even though he had no primary or general election opponents. One result of this conservative attitude among congressmen to their electoral fortunes was that political consultants could sell their wares regularly to safe incumbents, as well as to incumbents whose tenure was insecure and to challengers who had some chance of winning a seat.

## 2. Television

We suggested at the beginning of the chapter that the electoral geography of a district affects the value of television advertising as a campaign technique; the greater the number of electoral districts, and hence elected public officials, there are for any given level of public office within the viewing range of a television station, the greater the wasted audience for any candidate running for one of these offices. Moreover, this form of wastage is related to another difficulty facing candidates: when there are too many elections to which the local television stations can attend, it is difficult for any one candidate to gain much free publicity from television coverage of his campaign. This situation can be contrasted with an area where the television stations face only one or two elections for Congress, and a few more for the state legislature, and can give considerable attention to the campaigns, including sponsoring debates between candidates. Of the areas on which we focused, only Denver provided the congressional candidates with a low wastage of viewers: in the 1970s about two-thirds of the viewers of Denver's stations were resident in the city's congressional district.[26] Not surprisingly, television advertising was held by most local politicos to have been the key factor in the continued electoral success of Denver congresswoman Patricia Schroeder, and all emphasized the skill of her media consultant, Arnie Grossman, in producing convincing slogans and commercials. Schroeder was Grossman's first political client but, after her initial campaign in 1972, he greatly reduced his general advertising work to concentrate on media consultancy for election campaigns. His experience was the exact opposite to that of Campaigns West: widely acknowledged for his contribution to the 1972 victory, his business thrived both within the state and elsewhere. If television was a central element of congressional campaigning in Denver, however, and few disagreed with the Republican party official who claimed that television and radio advertising was 'absolutely mandatory',[27] it could not be used successfully to the exclusion of other campaign techniques. In this respect the Denver area was unlike New York City, where Ed Koch won the first Democratic mayoral primary in 1977 by relying almost exclusively on this medium; there were two main reasons for this.

In the first place, despite the widely held belief that elections were welcomed by television stations as sources of revenue, they were actually a mixed blessing for the stations. The price of peak-time political advertising slots was less than that for commercial slots. When the demand for commercial advertising was low, political advertising did provide an important source of income for the stations. When demand was high, however, they imposed a quota for the air-time which they would sell to political campaigns. The stations would not sell more than this to candidates, even if the latter offered to pay the commercial rate. This was because they feared it might lead to claims for 'equal time' by less well-funded candidates who would demand similar access, but who would not pay the higher rate. Whether the demand for advertising slots would exceed the supply in a particular locale depended on the prevailing advertising prices in that area and on the ability of candidates there to raise funds. In some parts of America in the late 1970s television stations were close to 'saturation point' with political advertising, while elsewhere they were not; Denver was an instance of the former. One Denver political consultant claimed '. . . in races in Ohio . . . it just never became a problem. No one was able to buy all [the time] that was available, but I have seen cases here in Colorado where candidates . . . have been able to buy more than is available.'[28] In contrast, in 1977 the New York stations, which had very high advertising rates, were faced only by municipal elections in New York City and state elections in New Jersey; they were not saturated by the demand from candidates and were able to provide Ed Koch with the air-time he wanted.

The second factor which helped to restrict the relative importance of television in Denver was tradition. On the one hand, this meant that most candidates were careful not to saturate the voters with advertising, and thereby risk alienating them. This point was emphasized by the campaign manager of one mayoral candidate who argued that television had been employed more in the 1975 mayoral election than the one in 1971, but that this growth would not continue in the 1979 election, because 'it was not a good idea to saturate TV'.[29] This manager claimed that, if after their campaign had reached its original fund-raising target, extra money became available, they might consider sending out an extra mailer but they would not exceed their original target on television advertising. On the other hand, tradition had produced an expectation among voters about direct contact with the candidate and his campaign workers. Denver consisted mainly of family houses, so that such a style of campaigning remained possible. In New York, though, three of New York's boroughs (Manhattan, Brooklyn and the Bronx) were dominated by multiple-unit dwellings. With increased crime, many apartment blocks in the city in the 1970s had security devices at their entrances, so that campaign workers could not gain access to the homes of a large proportion of potential voters. Obviously, personal campaigning by activists was still possible in many New York neighbourhoods, but in an increasing number it was not, and, whatever the expectations about appropriate campaign styles were, indirect methods had to be employed.

One of the main thrusts of our argument so far has been that the use of television in congressional campaigning necessarily varied from area to area, and that the factors which restricted its use normally limited its employment even more in state legislature and city council elections. Nevertheless, if in the three areas we examined there were no examples of state legislature and city council candidates who relied heavily on television advertising, there were several examples of some television campaigning being used by candidates at these levels of office. There were unusual circumstances in those cases where it was exploited, so they do not provide evidence of how television might be extended generally to offices of this kind. But, at the same time, they are indicative of how expansion in television advertising may result from attempts to overcome specific problems in campaigns; we discuss briefly two of these instances.

The first case was that of Beth Meader, an unsuccessful candidate at the special primary election in 1979 for the Assembly seat in the predominantly black district of west Oakland and Berkeley. Of course, at a special election some of the barriers to the use of television do not apply – in particular, there are no competing elections. Yet, curiously for a campaign that made more use of television in a state legislature campaign than any one previously in the East Bay, it was undertaken by a candidate who was supported by the more radical elements in the black community. The circumstances of the election were unusual, however, in providing an incentive for Meader to rely on television campaigning. The incumbent had delayed the announcement of his resignation, a move seen by many as benefiting his aide, and eventual successor, Elihu Harris. Meader's campaign had to be organized more hurriedly than was normal, and her position was made worse by the fact that she could raise considerably less funds than Harris. She lacked the money to attempt the district-wide mailings of literature which Harris was undertaking, and she lacked the time to run an activist-based campaign. Instead, she spent $20,000 on television advertisements, about one-seventh of the total campaign expenditures made by Harris, and a sum which would have brought her only about one and a half district-wide mailings.[30] Although in the sense we have identified they were wasteful, these peak viewing time advertisements did focus some public attention on her candidacy; it is arguable that she would not have got this publicity by any other means so soon before polling day. The strategy did not bring her victory, and it was seen subsequently by most political observers as proof of the conventional wisdom that television campaigning was futile in the East Bay. In contrast, Harris's campaign was seen as a vindication of the view that literature mailings were usually the key to success in the area. This view might be correct, but it should be noted that none of those who argued this actually claimed that Meader would have won, or would have come closer to winning, if she had spent her campaign money differently. Indeed, it could be argued that a television campaign was the only way she might have turned the election in her favour, given the advantages enjoyed by her principal opponent. In any event, the Meader case does illustrate how a candidate who would not normally have contemplated a television campaign turned to this when there were no rival elections, and when, ironically, a lack of finance and time prevented a more conventional campaign from being mounted.

The second instance of television being used for a specific purpose was in a city council election in Denver, when the incumbent spent a quarter of his $12,000 budget on ten advertisements. The candidate himself had been sceptical about the value of both television and radio advertising, but he had been persuaded by one of his advisers to buy the slots.[31] Having won the election with a relatively comfortable plurality, he believed his earlier doubts were misplaced, although he did not think it would have been effective to have devoted more campaign funds to this medium. The wastage was considerable, of course: only one voter in eleven in Denver was his constituent, and perhaps about a quarter of the viewers were not even resident in the city. Nevertheless, the strategy seemed defensible as an intelligent use of campaign funds for two reasons. Advertising rates in Denver were low in comparison with places such as New York: for his $3,000 the candidate was able to buy time on the much watched 'Good Morning America' programme as well as one slot on the Lawrence Welk programme on Saturday evening. He regarded this last purchase as important because of the high proportion of elderly white voters in one part of his district. Moreover, the candidate was no longer a young, active man who could conduct door-to-door campaigning in the district himself; television advertisements were not a substitute for this, but at least some voters were able to see him talking about the campaign, something which complemented the even more impersonal style of direct mail campaigning.

What these two cases show is that candidates for offices for which television campaigning was not usually thought appropriate might rightly believe in some circumstances that arguments about wastage should be overridden by others. This suggests that in the future there might be many more experiments by candidates for lesser offices, but that a distinction would still remain between major offices for which television campaigning was essential, and other offices for which it would be used only in specific circumstances. One problem with this conclusion is that by the late 1970s the structure of television was changing, and it is still not clear how much viewing patterns will change eventually, or how this will affect the use of television advertisements by even gubernatorial and US Senate candidates.

Until the late 1970s most television viewers watched programmes on local television stations; these were either affiliates of the three national networks (ABC, CBS, and NBC), which at peak hours carried nationally broadcast programmes together with local news programmes, or, in the larger metropolitan areas especially, were independent stations. This meant that, if a person was watching television, there were a relatively small number of stations he could be watching, so that it was fairly easy and inexpensive for the advertiser to make certain that his product came to the attention of most viewers during an evening. The advent of various systems of cable and satellite television broadcasting may alter this. In some areas the newer cable systems provide over one hundred different channels to the subscribers, some of which cater to special interests and do not carry advertisements; this represents a vast increase in the choice of viewing, even when compared to that available under the older cable systems, such as the ones in Manhattan. The long-term effects of these developments on the audience size of the network affiliates has been the subject of speculation, and it is by no means clear that a fragmentaton of viewing between a large number of channels will occur. To the extent that it does, however, it is likely to stifle the growth of television campaigning for all but presidential elections. If a large proportion of viewers are watching baseball games from out-of-state stations, or plays and films on commercial-free stations, then even a gubernatorial candidate will find that his ability to 'reach' potential voters through television has been restricted. Television became valuable to state-wide candidates because most stations in America had audiences within their state's boundaries, and it also became valuable to those congressional candidates whose districts largely coincided with the viewing range of local television stations. If this link between television audiences and a geographical region is broken, campaign strategies for major offices will change radically. At the same time, there will be much less incentive for lesser office candidates to make use of the medium for specific campaign purposes.

### 3. Radio

If television really 'came of age' as a political device in the 1960 presidential election, then it was through Roosevelt's fireside chats in 1933 that radio emerged as a major influence in American politics. Like television, radio in America consists of a large number of local stations, some of which are affiliates of the three national networks; usually, though, a city will have many more radio than television stations. Again, as with television, there is the potential problem of 'wasted' audiences, but radio advertisements are much cheaper than those on television. The result is that radio advertising can be afforded by many candidates for lesser offices; for example, Black Panther Elaine Brown, who by local standards had a highly financed campaign in the 1975 Oakland city elections, advertised on six radio stations but had no television advertisements.[32] Radio advertising is also less expensive because advertisements of a satisfactory standard can be produced without the skills of a consultant such as Arnie Grossman, whose expertise with visual images in a thirty second television slot can be critical for a campaign. Decisions about how much advertising time to buy on radio have been different from those about television, not just because it has been cheaper, but also because the markets have become more specialized. Increasingly, radio stations have directed their programmes at particular audiences: they play either rock, classical, country and western, black, or easy listening music or are stations specializing in news programmes. Peak-hour political advertising can be more easily directed at specific social groups than in the case of television. But because it is complex for campaign organizers to assess how best to exploit these different audiences, radio has rarely been used as the central medium in campaigns. More usually it is employed as a complement to other campaign techniques – whether that be television, direct mail, or door-to-door campaigning.

While radio advertisements were bought by many candidates for lesser offices in all three areas of our study, its use varied considerably within each area. Although, as Sabato argues, radio can have more impact than television advertising, because people are often alone when they are listening to it, there was no consensus among candidates and consultants as to whether it was indispensable, or how it could best help a campaign.[33] For example, one campaign manager for a major office incumbent in Colorado suggested that the value of radio was that it could be utilized to make negative remarks about opponents, a tactic which was inadvisable on television. On radio there was no danger of the tactic backfiring, because 'there is no visual image which lingers'.[34] That is, it provides a way of avoiding the kind of problem that plagued, among others, Byron Rogers's Republican rival in 1952 in taking advantage of opposition weakness. Again, a Republican party official in Denver claimed that, while radio had been brought in already in some Assembly districts in the city, it would have to be used more widely in the future as it became more difficult to contact people in their homes.[35] Yet among many campaign experts in Denver there was less enthusiasm about the possible uses of radio. One successful candidate for a city council district in Denver bought radio advertisements mainly to promote 'events' he was organizing in connection with his campaign, while another cited the 'wastage' argument and claimed that radio advertising was prohibitively expensive for reaching a small number of people.[36] Moroever, attempts to make frequently broadcast radio advertisements a central technique in a campaign have rarely been successful. For example, in 1979 one city council candidate, who was well known among Democratic activists, spent nearly $7,000 in this way and still failed to win one of the at-large seats.

Similarly, in the East Bay a conservative attitude towards the purchasing of radio advertisements was dominant. It was widely used in one Oakland city council election in 1973, but no candidate repeated the experiment on the same scale.[37] Perhaps the most commonly found strategy among candidates at all levels of office was to advertise on black radio stations, if their districts had an electorally significant black population. Often candidates did this even when they bought little advertising time on other stations. Although many of them claimed that they thought it was an effective technique, none of them was prepared to offer an explanation of this. Certain beliefs about black politics would not be articulated even in a confidential interview. The most plausible account of their behaviour was that the weaker links traditionally provided by political organizations in the black community had made many people unresponsive to campaign literature. This made it seem worth while to try any technique which might increase candidate contact with a large pool of potential voters. Yet radio was not always used extensively in attempts to mobilize blacks, even by black candidates. Some successful candidates employed it while others did not, and we do not have to turn to the leaders of anti-poverty-programme 'empires' to find instances where it was not used. Thus, in New York, while Major Owens bought a lot of radio time for his predominantly black Assembly district in Brooklyn, State Senator Carl McCall did not do so in a district which included part of Harlem.[38] (We discuss the effects of the anti-poverty-programme 'empires' on electoral politics in Chapter 8.)

Although the problem of 'wastage' will always be a barrier to a greater role for radio campaigning in contests for lesser offices, this has not been the only reason for its not being employed universally. Equally important has been a lack of knowledge about its impact on the electorate. In turn two factors seem to have been responsible for this. At least since the 1950s, its use in major office campaigns was always a secondary one, so that it became difficult to disentangle the impact of other techniques from that of radio. Furthermore, a natural conservatism among many candidates at the state legislature level discouraged experimentation, and very little accumulated wisdom has been acquired. Radio campaigning would seem to offer alternative strategies for the lesser office candidate who lacks the campaign workers for door-to-door campaigning. Indeed, in conjunction with direct mail, it would seem to offer opportunities for such a candidate which, clearly, television does not. Whether it does transform campaigning at this level will depend largely on knowledge about its efficacy in given circumstances being disseminated in the political community.

## 4. Direct Mail

Sending campaign literature through the US post office is a much older technique than either radio or television advertising. To take merely one example: in the 1940s East Bay congressman George Miller's family and volunteer workers put a considerable part of their campaign effort into placing literature in envelopes and addressing it to constituents.[39] With a list of registered voters, the funds to pay for printing and postage, and volunteers willing to address the envelopes, any candidate could campaign in this way. The Miller campaigns were simply following a long tradition – one dominant in areas where direct contact by campaign workers with potential voters was difficult or irregular. In California, where there were weak party organizations at the county level, no precinct organizations, and low-density housing even in cities, direct mail often became the major element of a campaign. However, technological advances in the twenty years after 1960 made some aspects of direct mail campaigning far more efficient.

Editing and revising the lists of recipients could be done very quickly on computers, which could also print letters and envelopes and even reproduce letters and signatures to give the appearance of a personal letter. These kinds of developments made direct mail in major office campaigns a capital-intensive, rather than a labour-intensive, operation, and the rapidly declining cost of computers increasingly brought this capital within the range of lesser office candidates. An even more interesting development was the growing link between fund-raising and campaigning. More efficient direct mail made it possible to raise funds by postal solicitations from targetted individuals – those who were thought to be the most likely contributors to a particular candidate or cause. This suggests a distinction between old and new styles of direct mail campaigning: the objective with the former was solely to disseminate information or propaganda, while with the latter it has the added purpose of raising money from like-minded individuals.[40]

In fact, this distinction is not that clear-cut. Particularly with a first mailing, a relatively unknown candidate for an office at, or below, the congressional level would primarily have been seeking publicity among political élites and sub-élites.[41] He needed this to build up campaign organization. His secondary aim would be to come close to covering his postal and printing costs, and at that stage he would not be concerned with providing himself with a large campaign fund. Of course, once he had publicized himself, he could then undertake subsequent mailings with an expanded list of recipients from whom he was concerned principally to receive donations. Moreover, while new techniques made a new style of direct mail possible, much of the literature which was sent in the 1970s was purely of the propaganda kind.

Unlike television advertising, which could simply be too expensive if used widely in a campaign and where there was often a problem of 'wastage', the new form of direct mail campaigning was accessible to many candidates for lesser offices. Providing he could obtain a reliable list of potential donors, this kind of candidate did not require the expensive computer facilities available to a Richard Viguerie; modern office equipment was all that was required.[42] The central problem with the technique was making out a convincing case for financial contributions to be made. The less important the office, the less likely it was that those outside the electoral district would consider a contribution worth while; the smaller the district, the less likely it was that the candidate would be making contact with potential donors whom he would not otherwise meet. Consequently, there was great variation in the utilization of the technique for lesser offices, and the contrast between Denver and the East Bay is an especially interesting one. In Denver the tradition of personal contact in political campaigning, together with the small size of state legislature and city council electoral districts, restricted the development of the technique; the candidate was expected to, and often could, make personal contact with potential donors. Furthermore, the structure of political offices in the Denver area, with only a few congressional seats and part-time state legislators, also limited its development, because there were relatively few sources from which lists could be obtained. Traditionally, direct mail had been used in the East Bay as the main technique of campaigning, and the new 'warlord' style of politics, which embraced both congressmen and state legislators, spawned the growth of lists of political donors. Even when a legislator would not give his own list to a local government candidate, from fear of further demands made on 'his' donors in the future, he might allow his own fund-raising expert to organize the mail solicitations for the candidate. This would preserve the secrecy the legislator wanted. Many of these confidential lists originated from the subscription lists of magazines and journals, and these could be purchased by anyone. After the first mailing they were continually modified by their owners in the light of experience. This suggests an additional problem facing the minor elected official. He might be too unimportant to sustain frequent solicitations by mail but, if it was not used regularly and revised, the list became an inefficient way of raising money. This provided a further incentive for co-operation with a 'warlord', thereby reinforcing the loose system of mutual obligations.

In the late 1970s the use of direct mail seems to have grown considerably in many parts of the country, despite increased postal rates. After 1960 the price of first-class postage in America increased more than most other prices: in 1960 the postage on a one-ounce letter was 4 cents, while by 1978 this had risen to 15 cents, an increase of 375 per cent at the same time as the consumer price index had increased by only about 225 per cent. This increased the problems facing a candidate in deciding how best to use direct mail for propaganda. The main choice facing him was whether to send literature by first-class or bulk-rate mail. The advantages of the former were that it increased greatly the probability that the literature would be delivered, and that it would be opened and the contents noted. The one advantage of the latter was its price; at the end of the 1970s a candidate who previously did not have a bulk-rate permit found this method became cheaper once he had sent 900 items, and with very large mailings his costs would be 50 per cent of those of a first-class mailing. (For candidates who already had a permit, any mailing was cheaper than first-class postage.)

Before discussing this choice further, however it is worth emphasizing that the switch to direct mail methods in the 1960s and 1970s was brought about by two interrelated forces. On the one hand, major office candidates had an increasing incentive to rely on their own campaigning and less on their party's efforts. Direct mail bypassed party campaign workers and literature 'drops'. On the other hand, this separation of candidate from party tended to draw activists away from party campaigns, made party campaigning less attractive, and forced lesser office candidates to become more self-reliant. Though individual campaigning was much less suitable for these offices, because of the expense involved, there was an incentive for the candidates to follow the example of the major office candidates. In cities such as Denver, where there was a sudden switch from party-campaigning to individual campaigning, the increase in the use of direct mail was striking.[43] Candidates who no longer were relying primarily on the fortunes of the party did not want to be involved in literature 'drops' any more: they did not want their literature being left in letter boxes with the literature of other candidates. They believed this reduced the impact of their own brochures and 'newspapers'. Consequently, even in Denver, most candidates planned on sending some direct mail communications, because at the outset of their campaigns it was impossible to know how many of their workers would be involved only in their campaigns and how many would also be assisting in other campaigns. While a major office candidate could attempt to attract campaign workers from outside the party organization, in order to overcome the problem of divided loyalties, this was more difficult for a lesser office candidate. Apart from his immediate friends, the sources of activists were limited, and knowledge of this would suggest a direct mail campaign rather than one based entirely on leaflet-dropping. Thus, there were pressures on candidates to switch from door-to-door campaigning and leafleting even at a time when mailing costs were rising more rapidly than the rate of inflation.

To put the matter rather crudely: in the late 1970s candidates had to spend more money (in real terms) than their predecessors of the

1950s to get a similarly competitive level of exposure in the electorate. Trying to get the voters' attention is an example of what Fred Hirsch called a 'positional good': if the voter is attending to candidate *A*'s campaign, he cannot be attending to anything else.[44] Since there is no evidence to suggest that voters are especially attentive to election campaigns anyway, to a great extent candidates must compete against each other for attention. For candidates, the advantage of party campaigning, of the kind found in Denver before the 1960s, was that in effect it placed a ceiling on the amount of campaigning it was worth while for the individual candidate to undertake for himself. His opponents, and candidates for other offices from his party, faced the same constraint. As with other positional goods, competition forced the individual candidate to send increasingly more communications just to receive the same level of attention as he did previously. Of course, this 'free-for-all' was not without some constraints. While the supply of political finance increased dramatically in the 1960s and 1970s, it was probably much more elastic for major office candidates than for lesser ones.[45] Together with rapidly rising postal charges, the result was that lesser office candidates found that, even though they were spending more from one election to another, they were often struggling to maintain effective exposure among voters.

In these conditions two of the most crucial questions facing a candidate were, To whom should he send campaign literature, and by what kind of mail? Particularly when the size of the electorate was large in relation to the importance of the office, the use of bulk mail and the targetting of key districts, or voters, were devices which had to be adopted. One state legislature candidate in New York, for whom there was no alternative to the use of the bulk-mail rate, argued that the main problem for any candidate using this service was making sure that the literature was not stored for months in corners of post offices.[46] He claimed that the problem provided at least one continuing campaign function for the city's Democratic clubs. Their members' personal contacts with Post Office employees helped the 'flow' of a candidate's mail – a task few candidates could complete successfully by using just their own contacts in post offices. However, getting the literature out of the post offices was only the first difficulty facing candidates. One legislative candidate in Denver had three mailings using the bulk-rate, but afterwards claimed that very little of it had actually reached the intended recipients.[47] In his district there were a large number of new apartment buildings, in which the turnover rate of tenants was high;

because bulk mail has never been redirected by the Post Office, few of his intended recipients ever saw his communications, and the new tenants rarely bothered to look at the leaflets because they were not addressed to them.

These sorts of difficulty notwithstanding, candidates often had no alternative to the use of bulk-rate, at least for some of their mailings, because of the expense involved in first-class mailing. For example, a congressional candidate with an electorate of 200,000 would have had to pay $30,000 in postage alone for one district-wide mailing in 1978, if he had used first-class mail. There were many ways of solving the cost-versus-effectiveness problem. One campaign consultant said that his practice was to use the bulk-rate for any mailings with more than 80,000 recipients, and to use first-class mail only in the last ten days of a campaign.[48] His solution to the problem of high tenant turnover was to continue using bulk-rate: he argued that it was wasteful sending first-class mailings to people who might have left the area entirely. However, he stressed that all decisions like this had to be taken within the context of the more general targetting strategy of the campaign. Targetting is crucial to direct mail campaigning, and it is to this we now turn.

Targetting was not new to the 1970s. Arousing your supporters, and doing little to antagonize the potential supporters of one's opponents, is perhaps the oldest electoral strategy of all. Moreover, since at least the mid-nineteenth century, American politicians have recognized that certain precincts in a city will be marginal ones and that particular attention will have to be given to them. In other words, political campaigning in America has never led to each registered voter receiving an equal share of a candidate's (or a party's) time and literature. However, there were two departures from traditional campaign practice in the 1970s which widened the gap between the most attended-to constituents and the least considered. Both these features were prominent in the campaigning of East Bay state legislator Bill Lockyer, and his is one of the examples upon which we draw. The first development was that, when it was thought appropriate, literature was targetted at specific individual voters, rather than at particular voting districts or precincts. Computer analysis made it possible to identify certain individuals as the best recipients, and this cut down the waste involved in sending mail to everyone in districts where there happened to be relatively large numbers of the type of people being sought. This sort of targetting could be used, for example, in the second of two elections held close together – such as a partisan general election held a few weeks after a primary, or a run-off election for a non-partisan office. Those who had voted in the first election would be prime targets for literature in the second election, because they would be the most likely voters. The second development was that, squeezed by the cost of mailings, candidates sought to minimize wasted mailings even when targetting at specific individuals was not possible. Districts with a history of low voter turnout came to be excluded from mailings, even when they were districts where the candidate could expect to win. New technology and financial pressures combined to skew the distribution of campaign literature: in one election Lockyer sent thirteen communications to some registered voters, while others received none.

Targetting became complex at the level of major offices in the 1970s. Computer analysis enabled more literature to be sent to those who were regular voters than to those who were not, and different leaflets could be sent to members of different ethnic or social groups. But in many elections for lesser offices the more detailed analyses could not be afforded by candidates. A typical example was an unsuccessful black candidate for the Oakland city council. He could afford neither computer analyses nor a city-wide mailing. Instead, he simply decided not to send any literature to the Oakland hills, where he felt there were few potential voters for him. He concentrated his literature on the 'flatlands' but, as he did not know who had voted for him in the primary, he had to send literature to every registered voter in the key precincts.[49] His was a good example of the continued use of old-style targetting. At the same election, though, a candidate for another council seat had greater campaign funds, and could make use of the newer techniques.[50] He discovered the 'voting histories' of individuals and these determined who would receive his literature; only regular voters got his more general literature, while there were smaller mailings to specific social groups. The former was sent to 48,000 people – about 14 per cent of the city's population – so that even in relatively well-financed campaigns only a small number of residents were receiving any literature.

The effect, then, of the new kinds of targetting were, almost certainly, to increase the amount of material received by regular voters, and to decrease that received by irregular voters. Of course, it cannot be demonstrated conclusively that this has contributed to decline in electoral turnout between 1960 and 1980, but it is a plausible hypothesis that it did so. If this was dysfunctional for American democracy, it stemmed from entirely rational behaviour on the part of individual candidates. If radio, and more especially television, helped to bring national politics into most homes, then the forces which helped to bring about the growth of direct mail at the local level also helped to take local politics out of many homes.

Yet, it might be asked, why do widely-held beliefs about the way electoral markets work not include some understanding of how the logic of the market can restrict the area of electoral conflict? The conventional view holds that, while institutionalized barriers to competition have been erected in periods of political upheaval, this is only because one-partyism has emerged. Essentially this is the view of, among others, E. E. Schattschneider.[51] The assumption is that, if the barriers are removed, the force of competition will always lead to the extension of the scope of conflict; he who thinks he faces defeat tries to mobilize more combatants. There are two limitations, however, to this model of the mechanism of electoral competition. First, it does not take account of the costs faced by candidates in mobilizing different sectors of the electoral market. Secondly, it assumes that a candidate or party will have a large group of partially-informed 'core' voters, and that primarily competition will be for the support of a smaller number of uncommitted or marginal voters. However, these 'core' groups have been declining as voters have become less loyal to their parties. When the 'core' is small, the candidate has to decide whether to allocate resources to persuading habitual voters, who are not completely committed to the opposing party or candidate, or whether to assign them to the persuasion of those who are less likely to vote. The issue of differential costs arises in that he has to motivate people in the latter group not merely to vote for *him*, but to vote at all. It is reasonable to suppose that this is the more expensive task, though ultimately it might lead to the conversion of more voters to his candidacy. It is worth while, then, extending a campaign beyond the groups of habitual voters only when it seems impossible to devise a strategy for winning a majority within them.[52] But to be done effectively, this will require a much more highly financed campaign. There are circumstances in which it is worth attempting this, but it is at least arguable that these are more uncommon than the circumstances in which opponents will prefer to fight for more habitual voters.

We would suggest, then, that there seems to have been a feature of the electoral market system itself which has probably helped to accentuate the decline in voter turnout in America's cities. Voters who had not turned out at the polls regularly came to have only limited contact with candidates, especially those for lesser offices, and this gave them less stimulus to vote at future elections. In brief the switch from party campaigning to high technology, direct mail campaigning may have helped to create 'two nations' in the electorate: those who received considerable attention from candidates and those who were ignored. Obviously, the effects of this change in campaigning must not be exaggerated, nor could we possibly claim that it is the main cause of turnout decline, or even that it is a major catalyst for it. But the direction of the change in candidate attention to different electorates is unmistakable, and it would seem implausible to claim that it made no contribution to declining voter turnout in the 1960s and 1970s.

## 5. Door-to-Door Campaigning

A variety of practices are covered by the term 'door-to-door' campaigning. In its most intensive form it consists of the candidate attempting to present himself directly to voters by appearing on their doorsteps. Apart from when it is done on a small scale for favourable publicity, it can be attempted seriously only in very small electoral districts. Even in an Assembly district in Colorado, a candidate would have to devote himself to the task for many months if he was to meet most of his electorate. One Assembly candidate interviewed in Denver in 1976 had quit his job at the end of the 1975–6 winter to campaign in this way; he was unsuccessful at the November election.[53] In the much larger districts of the California legislature, it would be impossible for even a challenger to meet more than a small proportion of the electorate in this way, and an incumbent would have the time to meet only a tiny number of voters. (In places such as New York, there is the added problem of gaining access to apartment buildings with security systems.) Another form of door-to-door campaigning, though one usually practised just before election day, is for the candidate to be 'on hand' to meet constituents in a neighbourhood, while his campaign workers ring the doorbells and inform the voters that they can meet the candidate if they wish to. This too is labour intensive but it uses up less of the candidate's time, and it is more 'personal' than his shaking hands outside a factory or supermarket. But with all forms of personal appearance by the candidate, there is the problem that he meets relatively few people and has less time to devote to other aspects of the campaign.

Traditionally, door-to-door canvassing by party workers on behalf of the candidate was an important aspect of campaigning in many cities. It was a device well suited to the densely populated cities, before the universal adoption of security locks on the front doors of apartment buildings, and one which was ill suited to the low-density housing of the suburbs. Indeed, it was not appropriate in many western cities, particularly when single family dwellings were located on hills – as in eastern Oakland and Berkeley. However, it was a method which could be organized efficiently by a party, because several candidates for public office were being represented by the single worker. Similarly it could be appropriate when candidates running for office in party primaries were organized into slates: a relatively small number of workers, in relation to the size of the electorate, was needed. In many cities party organizations survived for decades simply because the supply of campaign labour they could put on to the streets was in excess of that needed to make contact with most voters. These organizations experienced a decline in both their efficiency and their ability to recruit for patronage positions long before their demise as electoral intermediaries; they survived in a world lacking many alternative campaign techniques, and because once established the organizations did not need that many people to perform the electoral tasks for a party slate.

The party organizations were increasingly squeezed in two ways however. On the one side, there were newer techniques which made it possible for major office candidates to campaign more easily on their own. In becoming more independent, the candidates wanted to separate their canvassers from those of the party. Acting on the principle that a candidate usually gains another's opponents, and not his supporters, among voters when he is associated with him, candidates and managers have usually tried to recruit workers who were not involved in other campaigns. They did not really want activists who would 'push' other causes or candidates on the doorsteps, and they wanted people whose time was not being divided between several campaigns.[54] Inevitably, though, in many election years two or more candidates would be attracting the same activists. As we have seen, this rarely produced efforts to co-ordinate the campaign activities. One instance when some co-ordination was attempted was in the 1979 Oakland city elections, when two black candidates were seeking election to different seats.[55] They co-ordinated the distribution of some joint literature, but even in this case most of the literature promoted just the one candidate. On the other side, in the last forty years there have been developments which have made door-to-door canvassing more difficult and have required more campaign workers than previously. It was not a device which was very appropriate in the growing number of suburbs, and in the depopulating cities it became more difficult to canvass. Not only did buildings increasingly have security locks, but there were fewer people at home during the day. The rapid growth in the number of married women with jobs meant that there were more households at which canvassers would waste time discovering that no one was home. Canvassing became more of an activity for the early evening; with fewer hours in which to perform it, more campaign workers were required in the 1970s than twenty or thirty years earlier. This was at a time when activists were 'exiting' to the campaigns of individual candidates. The parties could no longer supply the workforce to mount effective door-to-door canvassing.

We have argued that the increased demand for workers by individual campaigns, combined with a decline in the ability of parties to mount effective campaigns of the old style, necessitated changes in campaign style. While major office candidates could choose between techniques, in the case of lesser offices candidates often had little alternative to the greater use of direct mail. Over several elections we might expect that these changes would lead to candidates adjusting their expectations and requirements, about the number (and type) of campaign workers appropriate for their campaigns. Thus, not surprisingly, while many of those interviewed spoke of a decline in the work-force regularly available for political campaigns, most candidates said that usually they did have enough workers for their own campaigns.[56] Yet, if the typical campaign of the 1970s was less likely to be linked to a party campaign, was promoted by fewer party workers and other activists, mailed more literature, and targetted the literature more specifically than in the 1950s, upsurges of activism could still generate a surfeit of workers. After all, the stimuli to amateur activism vary from one year to another. One of the outstanding instances of an upsurge was in the East Bay in 1970, when both the Dellums and Meade campaigns had so many volunteers – Dellums alone had over 2,000 – that the campaign managers had to invent jobs for them, so as to prevent the growth of disillusionment among the work-force.[57] But such instances became comparatively rare in the 1970s; as political conflict decreased, and as these kinds of challengers became incumbents, it was a period in which the campaign worker was gradually replaced by other techniques.

If the new campaign technologies helped to make major office candidates independent of their parties, and thereby helped to destroy the campaign functions of party organizations, it also had an impact on minor office competition for which much of it was too expensive. By the end of the 1970s the application of the cheaper forms of technology to lesser office contests was still in its infancy; deprived of a party, candidates often struggled to mount effective campaigns, and competition for votes tended to shrink to competition for 'prime' voters. If television and other devices helped to bring electoral competition for major offices into everyone's home, arguably it also helped indirectly to restrict the scope of competition at lower levels of office. Even where it could be practised, 'warlordism' was only a partial solution to the difficulties facing lesser office candidates, and for the most part restricting the scope of competition was their main hope.

## Notes

[1] As noted in Chapter 1, there was no major study of the growth of the new campaign technologies until the early 1980s. Among the best books on the subject are Sabato, *The Rise of Political Consultants*; Doris A. Graber, *Mass Media and American Politics* (Washington, DC: Congressional Quarterly Press, 1980); and Robert Agranoff (ed.), *The New Style in Election Campaigns*, 2nd edn. (Boston: Holbrook Press, 1976).

[2] Herbert E. Alexander, *Financing Politics*, 2nd edn. (Washington, DC: Congressional Quarterly Press, 1980), p. 9.

[3] Congressional Quarterly, *C. Q. Guide: Current American Government, Spring 1980* (Washington, DC: Congressional Quarterly Press, 1979), p. 61.

[4] Michael Malbin, 'Of Mountains and Molehills: PACs, Campaigns and Public Policy', in Michael Malbin (ed.), *Parties, Interest Groups and Campaign Finance Laws* (Washington, DC: American Enterprise Institute, 1980) Table 1.

[5] In response to the statement, 'Helping to raise funds for the Democratic *party* is an aspect of party work I particularly enjoy', 35 per cent of the precinct committee members 'mildly disagreed' and 43 per cent 'strongly disagreed' (n = 200). In response to the statement, 'Helping to raise funds for a Democratic *candidate* is an aspect of election activity I particularly enjoy', 39 per cent of the members 'mildly disagreed' and 30 per cent 'strongly disagreed' (n = 199). Not surprisingly, the members of the Executive Committee were somewhat more interested in fund-raising but among many of them it was an unpopular activity: in response to the first statement, 37 per cent 'mildly disagreed' and 38 per cent 'strongly disagreed' (n = 51), while 33 per cent 'mildly disagreed' with the second statement and 26 per cent 'strongly disagreed' with it (n = 52).

[6] Robert J. Huckshorn and John F. Bibby, 'State Politics in an Era of Political Change', in the American Assembly (ed.), *The Future of American Political Parties* (Englewood Cliffs, NJ: Prentice-Hall, 1982), pp. 95–6.

[7] Wattenberg, *The Decline of American Political Parties 1952–1980*, pp. 105–6.

[8] Interview no. 112.

[9] Interview no. 82. There have been few studies of campaigning, and even fewer comparative studies; the exceptions include David Leuthold, *Electioneering in a Democracy* (New York: Wiley, 1968) and Marjorie Randon Hershey, *The Making of Campaign Strategy* (Lexington, Mass.: DC Heath, 1974).

[10] Interview no. 72.

[11] In the case of some television stations, even this wastage may be high. For example, some Spanish language stations may have a high proportion of illegal immigrants among their voters. Moreover, there are a few stations near the Canadian border which are 'hooked into' the cable systems in Canada, and which have a larger viewing audience in Canada than in America.

[12] Congressional Quarterly, *C. Q. Guide*, p. 62.

[13] *New York Times*, 7 Sept. 1977.

[14] Interview no. 112.

[15] Interview no. 93.

[16] Interview no. 124.

[17] Interview no. 94.

[18] Interview nos. 100 and 101.

[19] *New York Times*, 24 Aug. 1982.

[20] Interview no. 124.

[21] Interview nos. 113 and 124.

[22] On the distinction between 'specialists' and 'generalists' see Sabato, *The Rise of Political Consultants*, pp. 8–9.

[23] Interview no. 113.

[24] Interview no. 112.

[25] In thirteen of these elections the incumbent faced a general election opponent.

[26] This was the estimate of one Denver-based consultant – interview no. 117. The boundaries of the Denver seat changed in 1972 when West Denver was removed from the district; in the 1982 redistricting this area was returned to the Denver seat.

[27] Interview no. 113.

[28] Interview no. 117.

[29] Interview no. 86.

[30] Interview no. 27.

[31] Interview no. 108.

[32] Interview no. 27.

[33] Sabato, *The Rise of Political Consultants*, p. 196.

[34] Interview no. 123.

[35] Interview no. 113.

[36] Interview nos. 92 and 93.

[37] Interview no. 26.

[38] Interview no. 133.

[39] Interview nos. 43, 44, and 45.

[40] For a discussion of direct mail see Sabato, *The Rise of Political Consultants*, ch. 4.

[41] This was the case with a campaign for a lesser office where the author was able to observe some aspects of campaign planning directly. The candidate had

access to the mailing list of an ex-congressman which he used to generate publicity at the early stages of the campaign.

[42] On Viguerie, see Sabato, *The Rise of Political Consultants*, ch. 4.

[43] This point was emphasized in interview no. 54 involving a long-serving senior party official in Denver.

[44] Fred Hirsch, *Social Limits to Growth* (London: Routledge and Kegan Paul, 1977).

[45] For example, while expenditure by state legislature candidates in Colorado increased by just over two and a half times between 1974, and 1978 expenditures by candidates for federal office more than trebled during the same period.

[46] Interview no. 133.

[47] Interview no. 94.

[48] Interview no. 124.

[49] Interview no. 68.

[50] Interview no. 72.

[51] Schattschneider, *The Semisovereign People*.

[52] Of course, there are circumstances in which activists mobilized around a particular issue provide a candidate with a source of campaign workers which enable him to extend the campaign to the irregular voters without any increase in cost. But these are the more unusual circumstances.

[53] Interview in 1975–6. This interview was not recorded on tape.

[54] We have described the use of this tactic in Ware, *The Logic of Party Democracy*, pp. 102–6, in the case of Senator Gary Hart. The manager of another candidate for major office emphasized that they used this tactic as well – interview no. 123.

[55] Interview no. 68. Of course, where slates have survived, as with the BCA in Berkeley, the problem of co-ordination does not arise in this way – literature drops are usually organized on behalf of the whole slate.

[56] However, the view of the candidates constrasts with the view of the Denver activists surveyed: 17 per cent 'strongly disagreed', and 43 per cent 'mildly disagreed' with the following statement, 'In most election campaigns in which I have been involved, it has not been too difficult to get a sufficient number of people to work in the campaign' (n = 251).

[57] Interview no. 51.

# 8
# Coalition Partners: Blacks and Labour Unions

The history of the Democratic party since the New Deal is closely related to the political emancipation of America's black population and to the rise, and subsequent slow decline, of the country's labour unions. For rather different reasons, it is appropriate to view them both more as allies of the Democratic party and its organizations, than as constituent parts of it. In the case of the unions, the rationale for this is perhaps more obvious. With the odd exception, such as Michigan, where the party was revitalized by the direct involvement of unions and union members in the party, most local unions and most local Labour Councils sought to maintain a political identity which was separate from that of the Democratic party. No attempt was made to claim that the Democratic party and the unions were parts of a wider 'labour movement', as with the relationship between the Labour party and the unions in Britain. The alliance forged by Roosevelt's policies was one of autonomous actors. In this sense, blacks could not be said simply to have been partners. After the New Deal blacks were the most loyal of Democratic voters; no major black political parties were established; and the Civil Rights movement was not like the labour unions – a regular financial contributor to partisan candidates, including some Republicans.[1] Of course, there were some cities in which northern blacks were integrated into an urban political machine in the 1920s or 1930s: Chicago's South Side is the best known instance of this. Yet blacks were not always embraced easily within existing party organizations, nor were many of them to be found within the ranks of the amateur movement of the 1950s, whether it was organized in clubs or involved direct participation in the formal party structures.[2] Indeed, in many cases black electorates and black community activists were accessible to white politicians only through various 'gatekeepers' – at first the church ministers, and later elected black officials and those who acquired posts in agencies of the Johnson anti-poverty programme – rather than through officials of the party organizations. Thus, unlike the longer established, white ethnic groups in the Democratic 'coalition', blacks were (and often saw themselves as) partners with the Democratic party and not a component of it. In this chapter we examine the changing natures of these two alliances, with first the black community and then the labour unions, to see how they relate to the other sources of party decline we have already considered.

## 1. The Unsteady Growth of Black Politics

Black migration from the south to the northern and western cities coincided with the early stages of party organization decline. It is worth remembering that as a black community, and perhaps the most famous black community in the world, Harlem is a twentieth-century phenomenon. Until the early years of this century, the black area of Manhattan was located in the Hell's Kitchen district of the West Side, while Harlem was white and middle class. Blacks were a small minority of almost all the non-southern cities at the time that Progressive influences in state Assemblies were legislating for party primaries and other anti-organization devices. Black political power came much later. For example, in Manhattan there was no black district leader in the party until 1935.[3] Thus, when we remember James Q. Wilson's contention that 'The Negro political organization is created and shaped by the political organization of the city', we must recognize that in few cities did blacks attain a position of any influence in powerful political machines.[4] As in many other respects, Chicago was an exception; there, a black was elected to the Congress as early as 1928, and Chicago was also peculiar, of course, in that the machine actually became more powerful after the mid-1950s. In that city wide-ranging and valuable patronage jobs remained in the hands of party leaders after the establishment of black communities, but elsewhere blacks became numerically important only as the manual jobs available to the parties had ceased to be as attractive as the jobs blacks could obtain without party assistance.

Together with one important feature of the 'southern experience', namely the centrality of religious ministers in the community, the timing of their migration brought about a very different relationship between blacks and the party organizations from that usual with the earlier, European migrants.[5] For the latter, the party provided a political structure, a hierarchy up which the faithful and useful could move. This is not to suggest that the parties were always eager to embrace new ethnic groups; one incentive for not doing so was that, for example, bringing in the Italians would reduce the number of jobs which could be distributed to newly-arrived Irish immigrants.[6] However, when they were numerically large enough, or when it seemed that the opposing party was close to building an electoral majority, self-interest dictated that they be included in the party's ethnic coalition and in its system of patronage. But this was a 'coalition' arrangement in which the party provided a framework (a county, district, and precinct organization) which would establish for the future who would constitute the political élite. Three factors were responsible for blacks not experiencing this process of integration into the party. Mass black migration to the urban north coincided with the era of one-partyism, of the 'system of 1896', so that urban parties were less subject to electoral pressures than they had been a generation earlier; racial prejudice was much more easily sustained against the descendants of black migrants than against any European group because they could not disguise their origins; and they brought with them a tradition in which ministers had quasi-political, as well as other, roles. Ministers were, and have remained, important 'gatekeepers' to the community, and consequently the party structures were a much less important regulator in determining who would emerge or remain as political leaders. If as Charles Hamilton has said of Adam Clayton Powell 'Because of the fragmented nature of New York City politics, Powell could maintain his Abyssinian [church] base pretty much independently of the city machine', it should be added that black churches were a major power base independent of the parties in many cities.[7]

There have been three principal effects on Democratic party politics of this interaction between party organizational decline and a community leadership centred on ministers of religion. The first is that black ministers have become candidates for elected public offices much more frequently than their white counterparts. Perhaps the most famous example of this phenomenon was Adam Clayton Powell who dominated Harlem politics in the 1940s and 1950s, and who was the leading black congressman in the 1950s and 1960s. He attained his political power from the success of his ministry. He followed his father as minister of the Abyssinian Baptist Church in 1937 – a church which already had 11,000 members and thirty-three staff at the time of his accession; eight years later he was elected to the US Congress. Despite the fact that the number of channels through which blacks gained public office greatly increased between the late 1940s and 1980, minister-candidates remained an important element in political recruitment. Of course, the rise of black public officials and other kinds of leaders reduced the advantages ministers enjoyed in seeking nomination and election, but their positions still helped their candidacies. The point is best made by comparing two such candidacies. Carl McCall found that he was able to emerge as a serious challenger to an incumbent state senator in New York in 1974 because he was a prominent minister in the district.[8] Yet at the same time, he had extensive support from both regular and insurgent factions in the district while the incumbent, who had been elected under unusual circumstances, had made little effort to reconcile himself with the activists. McCall's election that year could not have been achieved without that broader coalition of support. The second candidate was the Reverend Frank Pinkard who campaigned for an Oakland city council seat in 1979. With the exception of Oakland mayor Lionel Wilson, Pinkard's élite support came only from black ministers; he was opposed by the organizations and supporters of black politicians Ron Dellums and John George, as well as by almost all the white Democratic factions. Pinkard obtained 22 per cent of the vote, and failed in a three candidate contest to get into the run-off election. However, as one black political organizer, an opponent of Pinkard's, argued, the significant point about the result was that Pinkard had done very well, given that his support among the political élite was so narrow.[9] A similar point was made by a white Oakland councilman, who added that the importance of the ministerial alliance was also revealed in its endorsement of the winning candidate in the subsequent run-off election. To summarize the power of minister-candidates: to win they must build a coalition beyond the support of their fellow ministers, but even in defeat ministers acting in unison can be a major influence in determining who does win.

Linked to the power of ministers to act together in giving an endorsement to a candidate is a second factor which has affected the Democratic party. While there have been weak political organizations in the black communities, large captive audiences have been available in the churches, and this has made church services an important arena for electoral campaigning. While white liberal Democrats in the 1980s were becoming worried about the intertwining of politics and religion by right-wing fundamentalists, most forgot or ignored the fact that, for their black allies also, the separation of religion and politics had rarely been sustained. Many black ministers not only told their congregations for whom they should vote, but allowed candidates who had their support to address the congregations about the elections. This tradition had been maintained throughout the country, and the following examples illustrate the electoral role of the churches. In 1979 one black incumbent in Denver argued that the support of the ministers was critical in elections in his district, and he admitted that their support for him was made openly from pulpits.[10] In his mayoral campaign in Oakland thirty-two years earlier, Joseph Smith, who was white, was invited to speak in black churches, an invitation which was forthcoming through his labour union contacts; he later appeared before a meeting of the ministerial alliance to request their official endorsement.[11] Again, a former long-serving East Bay state legislator claimed that he had been able to get into any black church to speak on a Sunday morning, and that appearing before these captive audiences was the most productive form of campaigning.[12] That appearances before black congregations remained valuable in the 1970s is undeniable, what is more controversial is whether their value had declined.

On the one hand, there is the argument of those such as a former East Bay legislator who pointed to the much greater number of elected black officials whose own organizations provided a means of contacting voters.[13] Church appearances had become less critical because the kinds of campaigning practised by whites could now be undertaken by blacks. This argument is persuasive for the East Bay, when we compare the organizational resources available to D. G. Gibson in the 1950s with those possessed by Dellums, George, Wilson, Harris, and others in the late 1970s. Gibson's white allies had volunteers who could carry out door-to-door campaigning; but, in the absence of a comparable black middle class, there were few volunteers he could muster. Apart from the small organization of Byron Rumford himself, there was only Gibson's personal creation, the East Bay Democratic Club, which had far fewer active members than the white clubs. This was why Gibson became an intermediary between the churches and the white politicians – the church appearances he could arrange were a valuable asset for him which helped to compensate for his lack of manpower when dealing with white politicians. By the mid-1970s, with the collapse of the white club movement and with many more elected black officials, the gap in organizational strength between the white and black coalition members had been narrowed. There may still have been fewer black activists, and less finance for black candidates, but the huge imbalance in the relationship facing Gibson had disappeared. Black political organizations now existed; the churches which were once indispensable had now become one of several electoral resources in the black community.

Opposed to this is the view that the growth of alternative organizations did not reduce the value of black churches as providers of captive audiences for candidates. In spite of the new organizations, black voter turnout remained low in many cities and the most regular voters seemed to be church attenders. Of the low, and declining, level of black voter turnout there can be no doubt; for example, Charles Hamilton has shown how, in selected black districts in Harlem and Brooklyn, the turnout of registered voters in Democratic mayoral primaries was never more than 50 per cent of registered voters during the period 1961–77, and often it was much lower.[14] The relationship between church attendance and voting turnout was accepted by most of the black politicians we interviewed; but, as one of them noted, it was difficult to prove the strength of the relationship because the ministers kept their membership lists to themselves to maintain their own power *vis-à-vis* the politicians.[15] How valuable the black churches were in any given election seemed to depend on the following factors: the closeness of the contest, the candidate's other organizational bases, and his access to campaign funds. When these were not favourable to him, the candidate of the 1970s was much more like his predecessors in the 1950s in needing access to black congregations.

The continuing strategic position of ministers as 'gatekeepers' to the black electorate leads us to the third effect on the Democratic party: the sale of their support by ministers. Allegations of this have provoked bitter disputes between blacks and liberal Democrats. As with any aspect of politics which is regarded as corrupt, it is a difficult issue for the scholar to investigate. Throughout this research, allegations were made, especially by white politicians, that some black ministers solicited money for endorsements and for appearances before their congregations. In no instance, though, was it suggested that the ministers themselves benefited directly from the contributions. Moreover, on further probing of the issue, some of the allegations by interviewees turned out to involve the solicitation of funds to pay for the provision of campaign literature which the churches would provide for the candidate; what was really at stake here was loss of control by the candidate of some aspects of his campaign. Nevertheless, in some instances the politicians did believe that they were being asked to pay when there was no cost to be borne by the minister or the church, and allegations of this covered the entire period of our study.

In the East Bay there were suggestions that congressman Jeffery Cohelan had been faced by such demands, and more recently a row had developed there in the 1976 election between the Jimmy Carter organization and a group of ministers.[16] At the end of the 1970s there was a public uproar in Denver when a group of ministers solicited funds from candidates in a municipal election in return for their endorsement and a voter registration drive. The controversy discredited the plan and it was abandoned.[17] In New York the practice attracted little publicity because of the greater acceptability there of exchanging political support for material benefits. However, it must not be thought that the practice was approved by all ministers, or that all would employ it if they could do so without attracting adverse publicity. A relevant consideration here is that Christian denominations vary considerably, especially in the kind of training they provide their ministers. It was noticeable that those denominations which provided a more formal academic education for their trainees were the subject of far fewer allegations. In general it was the more fundamentalist denominations, with their rudimentary training, which were criticized; class differences, as much as cultural ones, seemed to underlie this source of antagonism within the Democratic party. Furthermore, white élite hostility seemed to stem as much from the belief that the black ministers were not expanding greatly the size of the active black electorate when they were being paid, as from middle-class scruples about payments for political support.

That much of the urban black electorate was not mobilized in the 1970s, and in many places was becoming less mobilized, is beyond dispute.[18] The vast increase in the number of non-southern black elected officials after the 1950s occurred in spite of low voter turnout, and there seem to have been three main causes for this increase: a rise in the relative, and often absolute, size of the urban black population; the continued concentration of this population in a few districts of a metropolitan area, so that voting power was not dispersed; and the Voting Rights Act of 1965 which made state legislatures and city councils less inclined to gerrymander electoral districts grossly to the disadvantage of blacks. In fact, these last two factors had a complex impact on the growth of black representation. Often areas were so overwhelmingly black that, while some black representation was assured, there were fewer black public officials overall than there would have been if districts which were 55 per cent or 60 per cent black could have been created. The very concentration of the black population meant that whites who wanted to restrict black voting power did not have to resort to oddly-shaped electoral districts.[19] With the rise of black public officials, political organization in black communities came to resemble more closely that found in white communities: one or two individuals, such as a D. G. Gibson or a Byron Rumford, could no longer monopolize the linkage between blacks and white Democrats. In place of the older arrangements, we find factional politics centred on individual politicians, especially where there was a large black population. In the areas we studied factionalism was most evident in the East Bay and Brooklyn. In Brooklyn it was so intense that sometimes a faction would back white opponents against the black candidates of the other faction.[20] Together with extremely low turnout in primary elections, this meant that in the 1970s blacks held fewer than one quarter of the seats in the two houses of the state legislature, and only one of the six congressional districts, in a borough where they formed 31 per cent of the population. In the East Bay the factions were less permanent and there was less impact on black office-holding. Even there, though, the failure of the black élite to back unanimously council seat challenger Elijah Turner in 1979 probably cost him the election; a similar split between the supporters of Harris and Meader in the primary election for what traditionally was the black Assembly seat in the East Bay almost led to the seat being won by a white candidate. In Denver there was no factionalism in the black community to weaken it as an electoral force *vis-à-vis* white Democrats; but there, neither alone nor in coalition with Chicanos, were blacks close to being able to run a candidate for any office above that of a seat on the city council or the state legislature.[21] Competition and division among elected black politicians weakened black power more than factionalism hurt white power, because, except in a few instances, black candidates were not successful in attracting white voters while black voters did defect to white candidates.[22] Quite simply, it can be argued that, to the extent it induced factionalism, electoral victories for blacks made it more difficult for other blacks to win elections in the future.

The consequences of public office centred factionalism in the black community were further exacerbated by unforeseen results of the Johnson administration's anti-poverty programmes. To understand this we must consider an argument outlined by Charles Hamilton.[23] Hamilton links low black voter turnout in New York to the fact that control of the anti-poverty programmes fell into the hands of a new type of black politician – one who was not involved in exchange relationships with 'clients'. If machine politics was characterized by an exchange between patron and client, then the new relationship was one between patron and recipient. Recipients provided nothing in return for the money and services they obtained, and there was no direct incentive for new patrons to politicize the beneficiaries:

> Recipients receive benefits . . . , but these are not a function of their voting . . . ; rather recipients receive these benefits through nonactivity and passivity or through episodic protest politics – another process. Thus this process is 'depoliticizing' precisely because it fails to take into account the very important fact that it is the *elective* political offices that ultimately control the fate of these 'soft money' programs.[24]

In this model of the political system electoral politics is divorced from the administration of public policy, and black electoral participation is thereby reduced. But there is another side to the matter which is of more concern to us: what happens to the black political élite when politics and administration are thus divorced? In this study we can do no more than hint at a possible answer to the question. One obvious point is that the introduction of a major new class – the elected and non-elected officials of the various anti-poverty agencies – into the political élite dispersed power. It restricted the flexibility of those in mainstream electoral politics and necessitated additional bargaining and coalition-building, and thereby power became more decentralized. There was a potentially more destructive alternative to this, however, when the programmes were a source of patronage to those who controlled them. Either to defend their position, or because they had wider political ambitions, these leaders tried to wrest political control of a district from the leaders in the electoral process, and thus monopolize power. This was disruptive because the poverty programme officials either had no direct stake in mobilizing a mass electorate, or it was secondary to their principal interests; their power more usually rested on low voter turnout in elections for Community Corporations.

In New York several politicians in the minority communities had established 'empires' in the anti-poverty programmes by the early 1970s, and two used their positions to mount serious challenges against senior members of Congress in 1976. In the South Bronx Ramon Velez became a rival to Herman Badillo, while in the Brownsville section of Brooklyn Samuel Wright posed a similar threat to Shirley Chisholm's incumbency.[25] Since we are focusing on blacks, we use the Wright case to illustrate the problems involved. Wright was a Democratic party district leader who was elected as a state assemblyman in 1965; at that stage his main political interests lay in the electoral process, but by the early 1970s this had changed. He ran various organizations connected with the anti-poverty programmes and was chairman of the community school board in the predominantly black and poor Ocean Hill–Brownsville district of Brooklyn. This latter base of his power emerged from the disastrous confrontation in 1968, which followed the teachers' union strike in the experimental system of community-controlled schools established there by Mayor Lindsay. Wright broke with the black activists who were on one side of the dispute, and after the teachers' strike had been settled, he publicized himself as the chairman of the school board who had rebuilt the school system.[26] Eventually, Wright came to control the annual dispersion of about $50m. of public funds. There were frequent allegations that his opponents were beaten up at school board meetings, however, and that he had profited from his positions: for example, it was claimed that he rented a building he owned to the anti-poverty organization of which he was chairman. Before his indictment in 1977, Wright had won 37 per cent of the vote in the congressional primary a few months earlier; this performance against a well-known incumbent was an indication of the power he had developed in Brooklyn in the 1970s.

In some respects Wright's practices differed little from usual machine politics; indeed, he had emerged from a machine and had been supported by the regular Brooklyn politicians and their judges; from 1970 onwards these judges removed, on technical grounds, fourteen of Wright's opponents from election ballots. However, Wright was not the creation of the Brooklyn machine nor was he controllable by it. His power base was independent of party chairman Meade Esposito, and theirs was an alliance of convenience; it was Wright, and not Esposito and the judges, who was primarily responsible for making his position safe against black critics. Just as all but one of the New York Democratic parties had been able to prevent court patronage from falling into the hands of reformers who could mobilize electoral majorities for major office elections, so Wright could protect his position in the poor electoral districts, with their low voter turnouts, in which he held his various offices. Nevertheless, it is the differences between the two sources of patronage which are crucial in understanding the problems for the politicians who were faced by an opponent such as Wright. Control of the courts provided the Democratic parties in New York with money from prospective judges, a judiciary sympathetic to the party when it was involved in litigation, and jobs for clubhouse members. In brief, it helped to maintain the basically labour-intensive campaign resources necessary for supporting candidates, and it existed to maximize political influence at all levels of office. By the early 1970s the regular organizations in New York were part of a now relatively stable system of activist politics. They and the reformers each controlled some districts, and shifts in control were gradual; in this situation elected public officials did not find themselves under continual threat from an unstable activist base. Furthermore, their position was being strengthened slowly as new campaign techniques reduced their dependence on activists.

In a place such as Brownsville the party activist base was smaller, and there were not the campaign funds needed for newer styles of campaigning. Wright posed a threat not just to liberals, but to moderate incumbents, because it appeared as if he might change the entire balance of political forces in the area. More significantly, though, Wright's interests were incompatible with expanding the power of the black community in Brooklyn or the city as a whole; his power bases were dependent on small electorates which he could control, and he could campaign for other offices for himself, or close allies, only when this did not involve greater electoral mobilization in his bases. Literally and figuratively, Wright's politics entailed a retreat into the ghetto. In his confrontation with Chisholm he found himself aligned against public office-holders who supported her more from fear of losing out to his politics of 'ghettoization', than from a commitment to her. Our argument, then, is that the anti-poverty programmes not only created 'recipients' rather than 'clients', but also could transform the black leadership into one of 'inward-looking patrons'. As with the kind of voter targetting we discussed in the last chapter, so here there seems to have been pressure for narrowing the scope of political conflict. Like targetting, this is an aspect of party collapse which other researchers might examine in much greater detail.

In addition to the electoral role of religious ministers and the political side-effects of the anti-poverty programmes, there was a more well-known feature of the recent history of black politics. This was the impact of radical politicians on black Democratic activists and leaders in the 1960s. Of the three areas in our study, this impact was greatest in the East Bay. In Denver the relatively small black population, and the large proportion of blacks who were middle class, minimized the problem. In New York the existence of an established black political structure within the framework of the Democratic party helped to reduce the potential for direct radical influence on electoral politics;

the Black Muslims remained aloof from this process in the 1960s, and there was no scope for independent Black Panther involvement in elections in the 1970s. In Oakland, however, there was a large black population which was under-represented in government, and a black political structure which centred on the Rumford-Gibson organizations and which had relatively weak roots in the Oakland ghettos. This was fertile territory for black militancy in the mid-1960s. Founded by Huey Newton and Bobby Seale in late 1966, the Black Panther party in Oakland had gained national attention by the following year.[27] Alongside white radicalism emanating from the student population in Berkeley, it made the East Bay the best known place in the country for radical activity. At that stage the Panthers throughout America were rejecting electoral politics, and in the wake of the collapse of the Rumford–Gibson axis – Rumford ran unsuccessfully for the state senate in 1966 – there was a political vacuum in the black community. The church ministers remained, but they had no standing with younger black activists who were attracted by radicalism and yet rejected the Panthers' methods. One serious alternative for them in 1968 was to join the state's anti-war white radicals in the newly formed Peace and Freedom party. Perhaps the critical decision which led to this option not being used was made by John George, who had chosen to run against incumbent congressman Jeffery Cohelan that year. Until then George's main political activities had been connected with the Civil Rights movement, so that his organizational ties to the Democratic party were weak. Nevertheless, he decided eventually to challenge Cohelan in the Democratic primary, rather than as a Peace and Freedom candidate at the general election, when his candidacy might have resulted in the election of a Republican.[28] By doing this he was able to broaden the anti-war coalition which had supported the white student candidate in the Democratic primary two years earlier, by allying them with younger blacks. Two years later Ron Dellums completed this process of coalition building in his victory over Cohelan.

If Dellums's election was a triumph for an army of political volunteers, essentially it was a victory for Berkeley political organization, where Dellums had become a city councillor in 1967, rather than evidence of a major new political organization in Oakland. In spite of later press references to 'Dellums's East Bay Machine', his ability to mobilize the black electorate in Oakland was, and remained, somewhat limited. Even with their informal nominating convention in 1971, the Democrats succeeded in electing only one candidate to the Oakland city council that year, and this was the first of several elections in which the limited electoral power of Dellums became apparent. There were two reasons why he could not emerge as the leader of a more united black community in the early 1970s. First, he was considered too radical by most black ministers; they had not played the part in recruiting him that they had in the candidacy of Byron Rumford in 1948.[29] This was to be a continuing problem for him. Secondly, although he was widely perceived as radical, and he maintained contacts with the Panther leadership, Dellums had little influence over that leadership. The real weakness of his position, and that of other Oakland Democrats, in relation to black electoral mobilization was revealed when Panther policy changed and they took up electioneering in place of militant posturing. In a campaign which also gained national attention for them, the Panthers decided to run Bobby Seale as a candidate in the 1973 mayoral election in Oakland. The penetration made by the Panthers into the otherwise ill-organized ghettos of west Oakland, and the weakness of other political organizations there, was shown when Seale finished second in the primary election to the incumbent Republican mayor. Certainly Seale was aided by Mayor Reading focusing attention on him in the non-partisan primary, and by the candidacies of both a white and a black liberal Democrat, but it is difficult to escape the conclusion that the Panthers had created a far larger 'grassroots' organization in the black community than anyone else had. It was the Panthers who had the volunteers to conduct door-to-door campaigning. Of course, they had little chance of winning for, as Reading knew in giving so much publicity to Seale in the primary, a large turnout of white voters could be guaranteed from fear that Seale might actually become mayor.

Thus we can see the irony of the Panther's role in undermining Democratic organization in the East Bay. While the spectre of armed militants might have been expected to have had an adverse effect on the party in the late 1960s, this damage was not great. It was when the Panthers turned to electoral politics themselves that the liberals suffered, for the Panthers could both out-organize them and also scare white voters into supporting conservatives. As one leading white politician in Oakland argued, the Panthers were important both for their campaigning and registration drive activities and also in mobilizing their opponents, and it was difficult to know which had the greater impact.[30]

That the Panthers could tap an unexploited source of electoral activism was revealed further in the two subsequent local elections in Oakland, in 1975 and 1977, when the party was no longer a threat to the Democrats' electoral ambitions. By 1975 the leadership of the Panthers had passed to Elaine Brown who herself ran for a seat on the city council that year. She had been unsuccessful in contesting a seat in 1973, but her second campaign involved a major effort to develop co-operation between the Panthers and Democratic organizatons, especially those associated with Dellums. As her campaign manager, Brown selected Beth Meader who, as we have seen, was later to run in a Democratic primary for the state Assembly. Meader, who had been active both in Shirley Chisholm's presidential campaign in 1972 and with the Dellums re-election campaign later that year, had never been a member of the Panther party. Moreover, her appointment was not the only substantive link with mainstream Democrats – for example, the Central Labor Council made available its phone banks to the Brown campaign, and the county party's Central Committee endorsed her. This campaign differed in two respects from the typical campaign liberals could mount in Oakland. First, it was exceptionally well funded: eventually more than $80,000 was spent, some of it raised through Brown's connections in Hollywood. Secondly, the campaign had the use of about fifty Panthers who undertook a variety of activities in contacting potential voters. Their main feature was the discipline with which they undertook their tasks, and this made them much more effective than similarly sized groups of other campaign workers. One person who followed the campaign closely described the Panthers in the following way:

How they achieved that discipline I don't know . . . but I suppose it was just the success of their propaganda, . . . or programme . . . that instilled all these people that had this motivation to do this kind of thing. You can't really be drugged to work day after day, walking from door to door, knocking on hostile people's doors.[31]

The Panthers had succeeded in doing what no other group had done in black Oakland – they had established a relatively large volunteer campaign organization which did not dissolve after an election. The result of this effort was that Brown obtained 11 per cent more of the vote than Seale had in the 1973 run-off election, but 41 per cent of the vote was still insufficient to defeat the incumbent.

The integration of the Panthers into an election coalition arrangement with Democrats continued in 1977. Once again about fifty

Panthers engaged in campaign activities, this time on behalf of Lionel Wilson in his bid to become the first black mayor of Oakland. A judge and a political moderate, as well as being the choice of all East Bay Democrats, Wilson could take advantage of the Panther labour force without harming his candidacy among white voters. Yet shortly after Wilson's victory, the first proof that the Panthers could make a significant contribution to a Democratic campaign, the Panthers ceased to have an electoral organization. What brought this about was the subject of much rumour, but what was undisputed was that Huey Newton returned to Oakland, Elaine Brown left the area shortly afterwards, and electoral activities ceased. At the end of the 1970s few political observers expected the Panthers to revive their large-scale electoral activities of a few years earlier. However, the Panther case belies the conventional wisdom that the political mainstream copies the methods of successful insurgent movements; mainstream Democrats could not find a body of dedicated activists to train in the methods of door-to-door campaigning practised by the Panthers, and black politics in the Oakland ghettos reverted to a state of under-organization.

This is not to argue that black Democratic political organization remained unaffected by the years of conflict, and then co-operation, with the Panthers. One consequence of the unsuccessful challenge in 1973 by black mayoral candidate Otho Green was the founding of the Niagara Movement Democratic Club. Until then there had been only two black clubs in the East Bay following the demise of Gibson's club, and neither of them attracted the younger, educated, politically more aggressive blacks who had been politicized in the 1960s. The Niagara club filled this void. It was similar to the older, and predominantly, white MGO club which was the last surviving Democratic club from the years of CDC influence and, as we saw in Chapter 5, they co-operated in the production of slate literature in 1979. We can thus see that in the East Bay black and white party politics had come to resemble each other more closely: there were Democratic clubs in both communities but they were weak in comparison with the elected public officials. There was no longer a single individual who controlled most white access to the black community. The main difference between the communities was that organization was essential to blacks if the full voting potential of the community was to be realized, and its absence greatly disadvantaged them.

To conclude this section. It is generally recognized that blacks were not as effective in the 1960s and 1970s as their population sizes in many cities would have indicated. The number of black public officials increased, but was still not close to reflecting their numbers in the population. Several factors outside black control contributed to this – the concentration of blacks in particular areas of cities, and continuing white hostility to black candidates, were just two of them. But we have tried to show that there were other factors which restricted the effectiveness of black political organization. The anti-poverty programmes gave patronage to blacks for the first time, but arguably this helped to de-politicize the community and to create conflicts between political leaders. Equally, while black militancy raised the consciousness of some ghetto blacks, it was only with difficulty that this could be harnessed to produce electoral pressures for change in the public policy process. Even when they were not restricted by powerful white organizations, as they were in Chicago, black politicians in many cities were not able to transform the role played by blacks in Democratic party politics; they remained very much junior partners in the alliance.

## 2. The Rise and Fall of the Labour Unions

During Franklin Roosevelt's administrations labour unionism in America prospered as it never had before. Except for a very brief period around 1920, the unions had never been able to claim as members more than about 10–12 per cent of those in non-agricultural employment.[32] By the end of the Second World War approximately 30 per cent of this work-force was unionized, reaching a peak of 32.7 per cent of the work-force in 1953. However, unionization never reached the levels it attained in many other western democracies, and the decline which set in after the early 1950s continued longer than it did in these other countries. By the mid-1970s just over a quarter of the work-force was unionized in America, and this was about half the 'union density' found in Britain or Australia and less than one third of the 'density' in Sweden. The long-term prospects for reversing this trend were poor. In the western states the decline in 'union density' was far higher than the national average: between 1953 and 1974 the median western state experienced a 10.4 per cent reduction in 'union density'; in no other region was the median reduction in 'density' greater than 4.8 per cent.[33] It was, of course, the west, and especially the south-west, which had become one of the main centres of economic expansion in the country. 'Union density' decline in the other area of expansion, the south, was low: in the median state 3.2 per cent less of the work-force was unionized in 1974. But, unionization was much lower in the south than elsewhere in 1953 – it was only half of the national average – so that the effect of even a small decline on union power in the south was considerable. Labour unions fared less badly in their traditional area of strength – the north-east; in New York 'union density' actually increased between 1953 and 1974, while the decline in the median state was only 3 per cent. But these were the states experiencing de-industrialization and low population growth, and which in some cases had a declining population.

The inability of American unions to recruit members in the expanding sectors of the economy can be shown in another way, by considering unionization among white collar workers. The proportion of union workers who were in white collar jobs did increase between 1956 and 1976 – from 13.6 per cent to 18.4 per cent of all union members. In Britain, however, the proportion of white collar workers grew much more rapidly than this in a period which overlaps with that for which American data are available: between 1951 and 1971 it increased from 23 per cent to 34 per cent of all union members. In other words, American unions were unsuccessful, certainly compared with their British counterparts, in making members out of the increasingly important white collar sector.

It is against this background that the relationship between the labour unions and its Democratic ally must be considered. In doing so, we focus primarily on changes in the two western states in our study. This is because the political effects of union decay can be seen there in an exaggerated form; the changes have been less spectacular in places such as New York City, and in examining Denver and the East Bay we can see more clearly the problems facing the Democratic party from its ally's collapse. It must be noted, though, that while the decline in 'union density' between 1953 and 1974 was higher in both Colorado (8.9 per cent) and California (7.5 per cent) than the national average (6.4 per cent), in neither state was it as high as in the median western state.

Writing two decades ago, when labour union power had only recently passed its peak, Greenstone said of the relationship between the unions and the Democratic party:

> Labor is unquestionably the most important and influential, and is usually the most reliable organized interest group supporting the party in national politics . . . This alliance is less pervasive in local politics.

Some relatively non-ideological unions, such as the building trades, are non-partisan, or may even support Republicans in some cities.[34]

Thus, while in presidential politics the union leadership was loyal to the party which had brought it into the national policy-making process, at other levels of electoral politics the AFL-CIO and the individual unions had a more pragmatic relationship with the party. Usually they supported whichever candidate had the best record on labour-related issues (or on issues directly relevant to their own unions) irrespective of party affiliation.[35] In perhaps one instance in ten they would support a Republican but the important point, as Greenstone's study of Detroit, Chicago, and Los Angeles demonstrated, was that the relationships varied from city to city. In the extreme case of New York some unions, most notably the Hatters' and the Ladies' Garment workers, chose to exert their influence through the state's Liberal party; indeed, the president of the United Hatters, Cap and Millinery Workers in the 1960s was also the leader of the Liberal party. The diversity of the relationships between Democratic parties and unions persisted, but in many cases the relationship changed as a result of five developments. The first was the continuing decline of party organization which we have examined in the earlier chapters of this book; the second was the decline in the proportion of the work-force that was unionized. The third development arose out of the reforms of the presidential nominating procedures; the unions lost their 'veto' over the candidates. In particular, the McGovern nomination in 1972 led to strains in the relationship between the party and the AFL-CIO leadership, and also between one union and another. A foruth development was a change in the balance of power in the union movement. Although unionization of white collar workers proceeded slowly in America, there was one sector of employment in which rapid unionization of a group which included white collar workers was occurring. These were employees in state and local government, of whom there were 4.1m. in 1952 but 13.4m. in 1980. 'Union density' among these employees increased by 1.1 per cent between 1964 and 1968, and by a further 6.9 per cent between 1970 and 1974.[36] Finally, there were the effects of the reform of campaign finance laws, at both the federal level and in many states, during the 1970s. These reforms reduced the relative importance of labour unions, *vis-à-vis* other groups, as donors of campaign funds; they also provided a disincentive to unions to be as active in giving money as they had been, because of the complexity of the required reporting procedures.

We examine the cumulative effects of these last four developments below, but, before doing so, we must outline the ways in which unions have been involved in campaigns for public office. At the possible risk of over-simplification, we would suggest that there were five levels of activity which we list in descending order of effort entailed:

(i) recruiting a candidate, or attempting to exercise veto power over the choice of the local party organization or the leading local public officials;

(ii) campaigning for a particular candidate or a slate of candidates through such methods as door-to-door canvassing of union households, sending their own campaign literature to union members, or arranging that a certain number of union members would perform electoral activities for a candidate during his campaign;

(iii) permitting a candidate's organization to use union resources – the 'in kind' contributions – such as printing facilities and telephones;

(iv) contributing money to candidates and party organizations;

(v) endorsing a candidate or a slate of candidates after they had been approached to do so.

Of course, in listing them in this way we are not suggesting that unions which practised (i) would always have practised (iii) or (iv), or that a union practising (iii) would also have practised (v). Even within one area the 'packages' of contributions different unions would supply varied. In Denver, for example, it was usual for candidates receiving 'in kind' contributions or money to receive the endorsement of the union, but one city council candidate, a Democrat, said that he had received money from the Teamsters but had not got an endorsement from them.[37] Moreover, while unions usually treated their endorsements seriously, for many candidates it was a relatively unimportant campaign contribution; as Bok and Dunlop argued: 'The effects of union endorsement cannot be measured precisely, but most professional politicians attach little significance to them.'[38]

Union involvement in elections in America has always varied from city to city, and it has been affected by such factors as the level of unionization, which unions were dominant in the area, and strength of party organizations. But, in most large non-southern cities in the years after Roosevelt's policies had made the unions part of the Washington establishment, they were active participants in local Democratic party politics. As Greenstone has shown, union participation in the nomination of Democratic candidates was considerable, even in those places where they could be outmanoeuvred by a party organization.[39] Furthermore, as we saw in Chapter 3, there were occasions when unions would even take the place of party in putting together a slate of candidates – this happened in Oakland in 1947. A decade later, when Jeffery Cohelan ran for the congressional seat in the Oakland–Berkeley area, unions were consulted at an early stage of his recruitment. After the 1940s and 1950s many local labour organizations (both individual unions and county labour councils) became much more passive in the nomination process, and the East Bay provides a good example of this. In part this transformation can be attributed to the further dispersion of power from formal party organizations, clubs, and local party notables to the candidates. Because these groups were less able to control nominations, irrespective of whatever agreements they reached with the unions, the latter would have to have become far more involved themselves in day-to-day electoral politics if they were to retain their influence. This became more difficult as the unions became weaker. In the more complex world of electoral politics in the 1970s, it was easier for unions to wait for candidates to approach them for endorsements and contributions, rather than play a more active role in candidate recruitment. By the late 1970s there was no evidence of organized labour in the East Bay having helped to recruit any candidates, and a revealing comparison can be made between Cohelan's recruitment in 1958 and that of Ken Meade in 1970. Both had strong connections with the unions, but in Meade's case labour's involvement in the campaign began only after he had decided to run for the state Assembly seat.[40] Even by 1970, the unions were withdrawing from a process in which previously they had been active.

In Colorado too union influence over nominations seems to have declined with the decrease in the control exercised by party groups over nominations. In the 1950s the leader of the teachers' union, Herrick Roth, was one of several legislators with union connections, and he was one of the most prominent state legislators from Denver; union influence over at least some nominations was assured because of party control in the large mult-member district. By the mid-1970s there was only one Democratic legislator from Denver who was closely connected with the union movement, and his links were with the relatively independent Building Trades Council. Roth's fate is also instructive. After becoming president of the state AFL-CIO, he became a candidate for the US Senate, receiving first-line designation at the state party assembly. Party organization and union support proved to be insufficient in the primary against Gary Hart, whose strategy was to bypass the party and appeal directly to the primary electorate.[41] As in the East Bay, no evidence could be found in Denver of recent instances in which unions had recruited candidates for the state legislature or city council, or even been instrumental in persuading a candidate to run, or withdraw, from an election to one of these offices. However, the unions still remained active endorsers of, and contributors to, candidates. The potential for continuing union involvement in the nomination process is obviously greater the more power the party organizations retain. Many traditional party organizations served the interests of some unions well, because usually they could rely on party officers to ensure that candidates acceptable to them were nominated. This form of centralization reduced the need for widespread union activity among incumbents, challengers, and potential challengers. When, as in Brooklyn and the Bronx, the regular organizations continued to control nominations in some districts, a few unions, and especially the Longshoremen, still chose to channel their campaign donations to candidates through the party.[42] Dealing with individual candidates was far more time-consuming, and resources could be preserved for conflicts with potential candidates hostile to their interests. Of course, this form of participation had never been open to unions in places such as the East Bay, because no party group there could determine who was nominated.

As with candidate recruitment, union participation in campaigning on behalf of candidates also appears to have become much less effective than it was. The causes of ticket-splitting and reduced party loyalty among the general population were just as potent among union families. The problem was not merely that unionists came to receive too many political communications, especially from television, to heed the advice of union leaders, it happened that in many cases labour leaders were having less contact with their members. *The East Bay Labor Journal* had been a weekly publication distributed free of charge to union households; by 1979 it was published only four times a year because the Central Labor Council (CLC) could afford it no longer. Political endorsements given in the editions published immediately before polling day had previously had a considerable impact, because they were contained in a newspaper which was read regularly, especially by the wives of members.[43] While the channels of information from the union leadership were being reduced in the East Bay, some of their other contributions to campaigns declined even more. The CLC election newsletter might still reach 10,000–15,000 people, but the number of precinct walkers the CLC could usually contribute to a candidate was no more than about ten by the end of the 1970s.[44] This small corps of volunteer campaign workers contrasts with George Miller's campaigns in the 1940s, when not only were there union volunteers, but there was also a number of members who were paid union rates by their unions to undertake door-to-door campaigning on Miller's behalf.[45]

In Denver there was an equally dramatic decline in organized labour's own campaign resources, though here there was disagreement on whether the unions distributed less literature of their own in the 1970s than they had in the 1950s.[46] In 1976 one union leader could still claim that the local unions were able to deliver the vote for the Democrats, and he cited as evidence of this the high poll for McGovern in 1972 in Denver districts which had a large number of union families.[47] But this aside, there was still little evidence that the unions were helping to mobilize voters. In 1978 the unions provided much of the campaign budget of $40,000, which was large by Denver standards, for a state Senate candidate whose opponent had supported 'right to work' legislation in the state. What they could not do was provide him with sufficient volunteers for him to distribute his weekly newsletter, the device which was central to his campaign strategy. Only on two days in the whole campaign did the unions send him enough volunteers, and he lost a seat which had been vacated by a retiring Democrat.[48] However, when only a handful of volunteers were required by a candidate, and when there were not many other candidates needing them, Denver unions could provide what was needed. Thus, one candidate for a city council seat was satisfied with the four volunteers he was sent on three occasions in his campaign.[49] But compared with most estimates of union volunteers participating in campaigns in Denver in the 1950s, this level of participation was very low.

The gradual disappearance of unions from the nomination process and from active campaigning was not matched, though, by a similar withdrawal in the provision of 'in kind' contributions to candidates. Passive participation of this sort was relatively cheap for the unions. For many candidates in the 1970s, the phone banks and printing facilities were the most useful type of aid which organized labour could provide, and there was no evidence that the unions had significantly reduced these contributions. In the East Bay the printing provided by the Operating Engineers was held by many candidates for lesser offices to be the most valuable of all contributions by unions. But, of course, the East Bay was an area of relatively low campaign budgets, and one in which direct mail was important, and the unions could not have provided comparable resources for radio and television campaigning. Furthermore, though their use was compatible with many campaign strategies, phone banks were normally made available only to candidates for the more important offices in the East Bay. While most candidates and campaign managers emphasized the difference between 'in kind' and financial contributions made by unions, there was one important respect in which their impact on union-candidate relations was similar. Often what was being provided was not an indispensable resource; money might be obtained by the candidate from elsewhere, just as he might be able to pay for a phone bank. Unions had influence with the candidate to the extent that he found fund-raising difficult or he was sympathetic to labour's causes. In contrast, thirty and forty years ago labour had power because its own campaigning among union households could 'deliver' the vote to a candidate. Obviously, only in a few districts was labour's support *essential* for victory, but finding a substitute for this element in his electoral coalition was more difficult for a candidate than finding alternative sources of money was to become in the 1970s.[50]

Before considering the unions' financial contributions to candidates and parties, it is worth recalling the point made by Greenstone in 1966 that in local politics some unions are non-partisan or may even support Republicans.[51] In 1972 leadership loyalty to the Democratic party became strained by the presidential nominating reforms and by the nomination of George McGovern. At the local level, though, it was the inter-union conflicts which were emerging at this time that were to strain the link between unions and the Democratic party. These were at least as damaging for the Democrats as rivalry between the AFL and the CIO had been in the 1930s and 1940s. In both the East Bay and Denver the prominence of certain environmental issues exacerbated the split between the building trades' unions and the county-wide labour councils – the Alameda County Central Labor Council and the Denver Labor Federation. Moreover, further division resulted from the shift in union membership – away from industrial unions and towards those in the growing service industries, including government itself. In Oakland there was in addition a physical dispersal of union headquarters which made swift, co-ordinated action less easy. Whereas in the 1950s many Easy Bay unions had their offices in the same building, with the officials having daily contact with each other, by the end of the 1970s the local headquarters were scattered around the city. The result of these difficulties in co-ordinating action, and of the withdrawal from the nominating process, is important; not only did unions remain inconsistent supporters of Democratic candidates at the local level, but even when they gave support they might give it ineffectively.

Whatever the lobbying power and organization of unions in Washington, at the state level and below much of their electoral activity was criticized by the candidates we interviewed.[52] This was particularly so in connection with financial contributions. Perhaps the most frequent comment made by candidates and campaign managers about the unions was that a candidate never knew how much money he would be given, and that much of it would arrive too late in the campaign to be used expediently. The most extreme instance of this concerned the Colorado state Senate candidate we discussed earlier. Not only was he not sent enough workers, but most of the money sent by the unions arrived too late in the campaign for his strategy and pattern of expenditures to be changed in ways commensurate with a highly financed campaign. A more typical victim of union ineffectiveness was an unsuccessful candidate for the Colorado state legislature who was less dependent on union money:

For some reason some of these unions hold the money back till like the third of November – the election being on the seventh. And in the final days we found ourselves with almost two thousand dollars of unspent money. So what we did was we took out fairly significant ads in the papers . . . and we [still] ended up with a balance of about a hundred dollars.[53]

His problem was that 'late money' for which no budgeting plans had been made could not be spent as judiciously as funds received (or, at least, known about) earlier in the campaign. In effect, inefficient methods of donation by unions reduced the value of the money they gave to a candidate, and thereby they reduced their future bargaining power with him. In an era in which money had become a more important political resource, in relation to others, many local unions still persisted with traditional methods of contributing, which were of little help to candidates. Yet this was only one of six problems that unions faced in their role as financial donors to Democratic candidates.

The second problem was that more expensive campaigns, conducted with some of the new technologies, were being fought at the very time that labour union membership was in decline. Not only did financial contributions not have the special character that being able to influence the votes of union families had, but, for unions to have maintained the influence they derived as the Democrats' largest financial contributors, they would have had to increase their gifts at the time that their main source of funds, membership, was shrinking. A third problem arose from the federal campaign finance reforms of 1974, which had the unintended effect of encouraging other organizations to distribute political funds in precisely the way that labour traditionally had. The reforms restricted unions' freedom in disbursing political funds, and at the same time they made donations by business corporations more efficient, because all donor groups were now required to comply with strict rules. Ironically, then, it was organized labour which suffered most from the disappearance of political slush funds. The rapid growth of business Political Action Committees was such that by 1980 the unions had lost their position as the largest group of donors to political candidates.[54] Corporate money became as important to many non-southern Democrats as union money, and the unions' bargaining power with these Democrats was thereby reduced. Fourthly, a further consequence of federal and state reforms of campaign finance was that the unions themselves could no longer 'launder' money. In states where unions were unpopular with the electorate this meant that candidates were now more likely to impose ceilings on the amounts they would accept from unions, because they feared their alleged dependence on organized labour might become an electoral issue. The manager of one successful Democratic candidate for state-wide office in Colorado claimed that his campaign organization had resolved that union money would form no more than 25 per cent of its total campaign funds. He admitted that while they 'could easily have got more' than this, they were worried about adverse publicity.[55] Finally, in some states, including California, new state regulations on the reporting of political contributions were seen as sufficiently complicated by union leaders as to provide a disincentive to making donations. While the leaders themselves denied this, many others we interviewed in the East Bay claimed that after the passing of legislation in 1974 there had been a partial withdrawal by the unions from campaign funding; it was argued that this was most noticeable in the case of elections to lesser offices.

Though their relationship with Democratic candidates was undermined in the ways we have just outlined, the unions' role as a financial contributor to Democrats remained a large one, at both national and state levels.[56] But the growing importance of money in labour unions' assistance to candidates weakened the old alliance with the Democratic party in a further respect. Money could be divided up quite easily in a bipartisan way; with most other campaign resources, it was more difficult to distribute effectively to both Democrats and Republicans, especially when party organizations were centrally involved in campaigning. In New York at the end of the 1970s an increasing proportion of union funds, particularly from New York City municipal unions, was going to upstate Republican senators. The reason for this change in the pattern of contributions was that the Republicans had usually controlled the state Senate in recent decades, and city unions believed they could acquire more influence there through co-operation, rather than conflict, with the Republicans.[57] That the municipal unions were the leaders in bipartisan funding is significant, because the growth of public sector unionism at state and local levels has been one of the most dramatic transformations in the composition of the labour movement.[58] It is a development which has had three principal consequences for the Democratic party.

First, unlike unions in the manufacturing, transport, or building sectors of the economy, government employee unions are disproportionately concerned with politics at the level of government from which they recruit their memberships. Of course, in highly centralized polities such as Britain, local government unions are necessarily concerned with the policies of the national government because these directly affect the interests of members. Moreover, their lack of autonomy forces local governments to co-operate with each other over issues such as pay and conditions of service. This is not so in America. Although, as a source of state and local government income, federal funds increased greatly after the early 1960s, federal government intervention was generally confined to setting uniform standards for the services to which it contributed financially. It did not interfere with the status of other levels of government as employers. This helped to discourage local affiliates of the unions from identifying with a nationwide 'industry', and instead encouraged the development of a view of the political world which was essentially locally-oriented. For a city employee, it was the government of the city which directly affected his livelihood; the influence of the state was usually, though not always, more indirect, and that of the federal government far less obvious still. Of course, in identifying this tendency, the difference between unionists in local government and in manufacturing industry should not be exaggerated. The tax revolts of the late 1970s, which began with Proposition 13 in California in 1978, and later Reagan's 'new federalism' provided state and local foci for local government unions.[59] Yet the central point remains: for state and local employees, the most important elections were those at the level of their employer. Compared with the 1950s, there were many instances in the 1960s and 1970s of local government employees being prominent participants in local elections, but they were not similarly active in state or federal elections. Two examples may be cited. In 1971 the Denver police, organized by their federation, held a big march during the mayoral campaign which was partly directed against the candidacy of liberal Dale Tooley; they were not similarly active in congressional or state legislature elections in 1970 or 1972. In New York municipal union support for John Lindsay in his 1969 re-election bid was extensive, and was a major reason for his unexpected victory; these unions were much less involved in non-municipal elections, except as financial donors. This non-national orientation of the unions is important, not merely because they have been one of the few expanding sectors of unionism, but also because their growth had influence on local labour councils. In this way the whole union movement was affected.

Secondly, in some cases the self-interest of government employee unions conflicted with the social policy objectives being promoted by the AFL-CIO and many non-southern Democrats. There have been many instances when this was not so but, as the expansion of the public sector slowed down or reversed, the strains became more evident. One problem was the failure of most governments to employ racial minorities in prestige jobs, such as in the police or fire services, and the subsequent demands by minority leaders for racial quotas in employment and promotion. There was also controversy over the residency of local government employees; many liberals argued that, in sensitive areas of contact with the public such as policing, those who did not live in a city did not have the same commitment to providing a high quality of service as did its residents.[60] Yet in many older cities a high proportion of local government employees lived in the surrounding suburban cities and counties. Naturally, they resisted attempts to change local ordinances to require employees to be residents, even when the change would affect only new employees. Another source of conflict centred on community involvement in the running of schools. Perhaps the best known of these problems has already been mentioned, that between the United Federation of Teachers and the experimental local governing boards of the Ocean Hill-Brownsville school district in Brooklyn.[61] This prompted a strike by teachers and, while the issue was a complex one, at its heart was the fact that it was a poor black district with a largely non-black corps of teachers. The switch from city control to partial community control exposed problems of racial conflict as well as those of the rights of employees. Here was an obvious demonstration of the complexity of the relationship between minorities and the labour movement which can be overlooked when examining national AFL–CIO support for civil rights.

Thirdly, in some cities, of which New York was the outstanding example, local government unions not only experienced a growth in membership but also in their bargaining power over employers.[62] Until the mid-1960s the mayor and his administration in New York had to contend with relatively quiescent unions, with one being played off against the others. As a result of reforms initiated by Mayor Wagner, seven unions became dominant and used industrial action to improve their conditions of service, especially their pension and retirement rights. For the electoral process there were two consequences of this. Unions and their new-found power became an issue that candidates could use against each other, especially when one was an incumbent – Edward Koch did this successfully in 1977. But the unions also became powerful actors in elections, using their endorsements as a bargaining device. As we have already mentioned, the Lindsay campaign of 1969 was characterized by the overwhelming support the municipal unions gave to the incumbent. The early years of his first administration had been ones of extremely poor relations with the unions, but a series of pay settlements altered this. In little less than a year Lindsay went from being a candidate without a winning coalition to victory, and the key to this transformation was the backing of the municipal unions.

The most significant point about such bargaining power was that it was incompatible with maximizing party control over the electoral system. The unions had interests which no longer overlapped with those of the party; they did not hope merely to be the beneficiaries of electoral victory by a party more favourable to organized labour, they bargained directly with incumbents, irrespective of party. That in 1969 Lindsay was a Republican whose early years in office were most notable for provoking conflict with the municipal unions was irrelevant to their stance, for they had been able to obtain concessions from his administration which made him a candidate superior to any that the Democrats could bring forward. Writing in 1970, Horton said of this new political position of the city's municipal unions:

The present status of unions stems not only from their own growth, but from the decline of New York City's Democratic party. The civil service unions, traditionally an appendage of the Democrats, now follow and reap the benefits of a more independent route. Potential mayoral candidates would as soon knock on union doors as on Tammany's for electoral support.[63]

Thus, when we find an exception to our earlier generalization about the partial withdrawal of labour unions from electoral politics, it is not an instance of union activity strengthening the Democratic party, but one growing out of party decay and one which helped to destroy the Democratic party further. The growth of municipal unionism, and the demise of industrial unionism, had inevitably weakened the relationships between local unions and Democratic parties. Like the alliance with blacks, the ties between organized labour and the Democratic party had become more complex. What we must not do, as regrettably too many writers on the subject do, is to over-emphasize the role of party decay and national party reforms in producing this. The decline of the party organizations was important, but what should have become clear by now is that there were several other factors responsible for transforming the links with organized labour and the role of the party in the electoral process. It is only by taking account of these that we can explain why party decline should have happened so quickly and at the time that it did.

## Notes

[1] On the loyalty of the black electorate see Robert Axelrod, 'Where the Votes Come From: An Analysis of Electoral Coalitions, 1952–1968', *American Political Science Review*, 66 (1972), 11–20.

[2] On the absence of blacks from the clubs, see Wilson, *The Amateur Democrat*, ch. 9.

[3] James Q. Wilson, 'Negro Politics in the North', in Harry A. Bailey, Jr. (ed.), *Negro Politics in America* (Columbus: Merrill, 1967), p. 318.

[4] Ibid., p. 316.

[5] '. . . the white church is but one of many white institutions. On the other hand, the black church was one of the few black institutions left relatively free by whites in its development and modification.' Hart M. Nelsen and Anne Kusener Nelsen, *Black Church in the Sixties* (Lexington: University of Kentucky Press, 1975), p. 1. See also Charles V. Hamilton, *The Black Preacher in America* (New York: Morrow, 1978).

[6] Not only did the old immigrants not want the new arrivals in the party, they did not want them in the electorate either. As Martin Shefter has observed, '. . . the predominantly Irish leadership of New York's Democratic machines, and

their predominantly white Anglo-Saxon Protestant counterparts in the Republican party, had little interest in sharing power with the immigrants from Southern and Eastern Europe . . . Hence, these party leaders did not make an all-out effort to bring the members of these groups into the electorate.' 'Regional Receptivity to Reform: The Legacy of the Progressive Era' pp. 479–80.

[7] Hamilton, *The Black Preacher in America*, p. 112.

[8] Interview no. 133.

[9] Interview no. 9.

[10] Interview no. 92.

[11] Interview no. 12.

[12] Interview no. 16. Ministerial alliances also sent out campaign literature to their congregations at most elections, thus informing the church members whom the ministers had endorsed.

[13] Interview no. 16.

[14] Charles V. Hamilton, 'The Patron–Recipient Relationship and Minority Politics in New York', *Political Science Quarterly*, 94 (1979), 220.

[15] Interview no. 130.

[16] Interview no. 32.

[17] Interview nos. 92 and 112.

[18] In early 1981 the low turnout of blacks in New York City was cited as one of the causes for blacks being in their weakest political position in the city for about twenty-five years; *New York Times*, 29 Mar. 1981. There were two elements of this apparent weakness. First, after 1977 there were no black members of the Board of Estimate, and hence no major elected officials at the municipal level. Secondly, it was widely believed that black interests did not count for much in policy-making under city mayor Ed Koch.

[19] Ironically, it is now the need to conform with the Voting Rights Act which requires the drawing of oddly shaped districts; this was the case with the 1982 redistricting in New York City.

[20] *New York Times*, 9 Oct. 1980.

[21] Colorado did have a black Democrat serving as Lieutenant–Governor between 1974 and 1978, but he was nominated only because of the efforts of white liberals. In 1978 he was dropped from the ticket when many of his white supporters abandoned him – a decision which provoked considerable dissatisfaction among black politicians.

[22] For example, Hamilton has noted that black voters have not been especially loyal to black candidates in mayoral elections in New York City in 'The Patron–Recipient Relationship and Minority Politics in New York'.

[23] Hamilton, 'The Patron–Recipient Relationship and Minority Politics in New York'.

[24] Ibid., pp. 217–18.

[25] In addition to Wright's organization, State Senator Stewart also developed a position of power in Brooklyn through the anti-poverty and urban renewal programmes there. See *New York Times*, 20 June 1970 for an account of this issue in a contentious primary election.

[26] *New York Times*, 9 Mar. 1971.

[27] A major article appeared in the *New York Times* on 6 Aug. 1967.

[28] Information obtained in interview no. 21, from a source close to John George.

[29] Interview no. 16.

[30] Interview no. 56.

[31] Interview no. 27.

[32] George Sayers Bain and Robert Price, *Profiles of Union Growth* (Oxford: Basil Blackwell, 1980), ch. 3.

[33] These figures were recalculated from the data presented in Bain and Price, *Profiles of Union Growth*, p. 92. In the three other regions the declines in union density were, respectively: mid-west 4.8 per cent, south 3.2 per cent, north-east 3 per cent. (Alaska and Hawaii are excluded from the calculations.)

[34] J. David Greenstone, 'Party Pressure on Organized Labor in Three Cities', in M. Kent Jennings and L. Harmon Zeigler (eds.), *The Electoral Process* (Englewood Cliffs, NJ: Prentice-Hall, 1966), pp. 57, 79.

[35] Derek C. Bok and John T. Dunlop, *Labor and the American Community* (New York: Simon and Schuster, 1970), pp. 396–7.

[36] Bain and Price, *Profiles of Union Growth*, pp. 96–100. A redefinition, by the Bureau of Labor Statistics, in 1968 of what constitutes a labour union makes it impossible to provide continuous data for this period.

[37] Interview nos. 112 and 93.

[38] Bok and Dunlop, *Labor and the American Community*, p. 413. Endorsements become more important the lower the level of office because they help a candidate build up credibility in the run-up to the election. Moreover, endorsements are taken very seriously by those making them. The East Bay COPE file for the 1974 elections reveals that, on 18 and 19 September, COPE's interviewing committee met forty-four candidates for fifteen minutes each. All the candidates were seeking election to one of four public boards. For each office at least eight standard questions were asked of the candidates.

[39] Greenstone, 'Party Pressure on Organized Labor in Three Cities'.

[40] Interview no. 7.

[41] Ware, *The Logic of Party Democracy*, pp. 102–6.

[42] Interview no. 133.

[43] Interview no. 49.

[44] Interview no. 27.

[45] Interview nos. 43, 44, and 45.

[46] Interview no. 126.

[47] Interview, Aug. 1976.

[48] Interview nos. 100 and 106.

[49] Interview no. 93.

[50] That congressmen are usually able to decide which coalition of interests to base their re-election bids on is emphasized in the work of Lewis A. Dexter. See, for example, *The Sociology and Politics of Congress* (Chicago: Rand McNally, 1969).

[51] Greenstone, 'Party Pressure on Organized Labor in Three Cities', p. 79.

[52] For a general account of American unionism in national politics in the 1970s see Graham K. Wilson, *Unions in American National Politics* (London: Macmillan, 1979).

[53] Interview no. 90.

[54] 'Though labor union PACs have steadily increased the amounts of money they have raised and spent on behalf of favored candidates, the percentage of increases (tripled from 1974 to 1976) by the corporate and business-related (including trade association) groups far outstrips that of labor groups (up 30 per cent in the same period).' Alexander, *Financing Politics*, p. 86. Alexander further shows that contributions by special interests rose from $8.5m. in 1972 to $35m. in 1978.

[55] Interview no. 123.

[56] See Alexander, *Financing Politics*, especially chs. 4 and 7.

[57] *New York Times*, 14 May 1979.

[58] As we argue below, accompanying this growth in unionization there has also been an increase in conflict between employers and labour unions, which has taken the form of well-publicized strikes by different kinds of employees. Examples of these strikes in the late 1960s are given in David T. Stanley, *Managing Local Government under Union Pressure* (Washington, DC: Brookings, 1972), p. 3.

[59] On the tax revolt see Robert Kuttner, *Revolt of the Haves* (New York: Simon and Schuster, 1980).

[60] This point was made in an informal discussion after an interview by a leading Democrat in Denver, a middle-class man who chose to live in a middle-class district close to the city centre. When the police arrived after he had reported a burglary, they were far from sympathetic about his misfortune. As residents of a neighbouring suburban county, they could not understand why someone like him would want to live in what they described as 'the jungle'.

[61] Morris, *The Cost of Good Intentions*, pp. 108–16.

[62] Raymond D. Horton, 'Municipal Labor Relations in New York City', in Robert H. Connery and William V. Farr, *Unionization of Municipal Employees* (New York: the Academy of Political and Social Science, 1970).

[63] Ibid., p. 77.

# 9
# Conclusions

There can be little doubt that what happened to the Democratic parties in America between the early 1960s and the late 1970s was truly extraordinary. Within a few years most of them were transformed, with the individual candidate for public office becoming the dominant element in the parties. In the preceding chapters we have tried to identify the factors which brought this about in three very different urban areas. Now we must draw together the different strands of argument in order to consider the question, Why did this transformation occur? In doing this, it is perhaps most useful if we attempt to classify the numerous factors which seem to have contributed to party collapse. One way of doing this is to distinguish between (i) factors which have facilitated candidate independence; (ii) long-term forces in American society and politics which had an impact on the party organizations; and (iii) factors specific to the political upheavals of the 1960s.

The first category is the most obvious; it includes the development of new technologies which could be employed in political campaigning, and the resources which helped incumbents, especially legislators, to divorce themselves from the nominating and electoral activities of their party's organizations. The second category includes several different factors which would have affected the parties, irrespective of the state of campaign technology or of political controversies in the 1960s. There were demographic changes in the cities which made the various non-white groups, especially blacks, more numerous, and these were groups which had been less well integrated into the Democratic party. Again, there were declining resources available for attracting professional activists into the parties and difficulties in providing solidary incentives to amateur activists, to complement the purposive incentives. Increased party competition at the state level, and other agents of what we have called 'democratic modernization', provided the impetus for reforms, such as single-member districts, which restricted party control. Finally, there was the declining strength and commitment of the Democrats' labour union allies. The third category includes a number of forces which seem to have been the direct product of the convulsions of the 1960s – intra-party divisions and increased party disloyalty; issue extremism; the opportunities for 'exit' to non-party political activity; and the institutional reforms which emanated from the conflicts, such as the reform of the presidential selection process. Obviously, several factors might be fitted into more than one of these categories. For example, are the district-servicing facilities legislators voted themselves better seen as an element of candidate autonomy, and the product largely of candidate independence, or as an aspect of 'democratic modernization'? This issue can be left, since we are concerned with discussing general trends, rather than devising mutually exclusive categories for the different elements.

We begin with the new campaign technologies, and the consultants who have sold them to the candidates, for, as we have seen, scholars such as Larry Sabato have refused to accept that the consultants were responsible for the demise of the parties.[1] Of course, we cannot be sure what would have happened if the other social, economic, and political changes of the 1950s and 1960s had not come about, but it is not that difficult to imagine the impact of the campaign technologies in the absence of these changes. The new techniques and strategies would have been as attractive to many major office candidates in a hypothetical 1960s and 1970s as they were in reality. As in the real world, 'outsider' candidates would have led the way in their adoption, and they would have been followed by some candidates who were more closely allied to their local and state parties. To return to the analogy we introduced in the first chapter: it is implausible to argue that industries would not have turned to the railways for transportation in the 1830s even if most British entrepreneurs had been satisfied with the canal system. Initial caution by candidates who were already semi-independent of their parties would have been replaced by a desire not to lose out to competitors who had successfully 'taken a chance' with the new services; successful experiments would have been copied elsewhere. The evidence from the 1950s is that major office candidates were experimenting at that time with the new technologies and, although this was not always fruitful, there is no evidence to suggest that the consultancy industry could ever have been stillborn. On the contrary, the decentralization of the caucus-cadre parties, compared with modern European parties, together with the absence of effective legal constraints on campaign expenditures, provided ideal conditions for individual candidates to experiment with the consultants' wares. To argue that the British case shows that the development was not inevitable in America is to ignore the fundamental differences between party politics in the two countries. In Britain party centralization, and a tight legal control over election expenditures dating back to 1885, provided no opportunities for candidates to separate themselves from parties by employing consultants.

That many lesser office candidates would be disadvantaged by a switch from party-centred campaigning would not have slowed down the process, and experience seems to show that it did not. At least initially, the campaign consultants had little to offer these candidates, but the latter had no say in the growing independence of the major office candidates. As those contesting major offices made a decreasing relative contribution to party organization election campaigns, so candidates contesting lesser offices had to provide more campaign resources for themselves. Nevertheless, it is not unreasonable to suggest that, without, for example, the demographic changes which took place in the cities, the separation of lesser office candidates from their parties would have occurred much more slowly. In short, the technological revolution of the post-war years would not have destroyed party organization control over state legislators and city councillors in our hypothetical world as quickly as it did in the real world. Yet, eventually, the erosion of control over governorships and congressional seats must surely have undermined the viability of city party organizations. The loss of patronage and influence over the public policy process at the state and federal levels would have isolated local politics in a way that it had not been isolated since the early twentieth century. But this isolation could, surely, not have been complete: some individual congressmen may well have found it to their advantage to practise something akin to the 'warlordism' we found in California, and thereby have threatened local organizational monopolies. We would contend, therefore, that it seems implausible to deny that the rise of the consultants was a sufficient condition for the ultimate collapse of local party organizations, even though there were other factors which acted as catalysts for the rapid collapse which actually occurred in the 1960s.

But was their rise a *necessary* condition for this collapse? Let us now try to envisage a political world in which campaign technology remained as it was in the mid-1940s, but where the other social, economic, and political changes of the next two decades proceeded as they did in the real world. Let us also exclude the consequences of the issue conflicts of the 1960s. That is, we are assuming that America experienced a change in the racial composition of its cities, a rise and subsequent fall in middle-class political activism, a decline in labour unionism, court-ordered redistricting, and a shift in the kinds of patronage available at the city level. In this hypothetical world major office candidates could not use television campaigning or sophisticated direct mail and polling techniques; they would have to rely on door-to-door campaigning, leafleting, billboards, and personal handling of direct mail. In one obvious respect, therefore, there would be a much weaker incentive for these candidates to loosen their ties to the party organizations than was actually the case in the 1960s. Yet compared with 1946, these party organizations would have much less hold over their members in the precinct organizations, less assistance from labour unions, would face stronger leaders in the black communities, and might now have the disadvantage of having to fight in single-member districts in state legislature elections. Any candidate who could gather together an army of personal supporters, and who could raise the money to advertise in newspapers and send 'mailers', would find exemplars of individualistic campaigning whom he could copy. In suitable congressional districts, for example, candidates would have the model of a John Carroll, Byron Rogers's predecessor in the late 1940s, to follow rather than the party-oriented model of Rogers (see Chapter 3, note 35). Moreover, the experience of California in the 1930s and early 1940s had shown that an individualistic style of campaigning was possible at all levels of office, even when modern campaign techniques were in their infancy. What is at issue, then, is not the possibility of some forms of individualistic campaigning, but the incentive to move over to that style.

There are at least two main types of response by candidates which we might envisage in a world with the electoral technology of the 1940s. On the one hand, we must consider the case of a now relatively inefficient party organization competing against another of its kind; both parties can still provide basic electoral services, albeit less extensively than before. We might expect candidates to rely as much on their party's campaign efforts as their predecessors in the last few decades had done. A slow decline in the quality of the services provided by both parties would probably produce a generally conservative response by candidates; they would continue to work with the party organizations because they would not see themselves threatened by changed circumstances, and because accumulating one's own campaign resources would have been far more difficult with the campaign technology of 1946 than with the technology actually available in the early 1970s. Indeed, the evidence from this study and those of other cities in the period we described as the 'Indian Summer' for the parties, the 1940s to the early 1960s, suggests that this was probably a fairly common response. Only in the last few years of the period were the technologies being used widely, and, in spite of the apparent growing inefficiency of many urban machines, this was not a period in which there was a major shift in the balance between party and individual campaigning. Generally, candidates seem to have put up with party organizations whose manpower resources were declining.

There are on the other hand conditions in which a more radical response from the candidates would be more likely to be forthcoming – when they would choose to depend on campaign organizations of their own creation. In particular, there are three conditions in which we would expect an individualistic style to be adopted: when there was a sudden and major decline in the electoral resources available to the party; when the party had reached the point where it lacked the resources to provide an adequate campaign at all; and when a candidate's own party was at a considerable disadvantage in its resources compared with his opponent's party. The experience in Denver in the mid-1960s was a good example of the radical response; there neither the new technologies nor the 1960s issue conflicts had much effect on the separation of candidate and party. After the adoption of single-member districts, candidates realized that the county party no longer controlled the nomination process and could not coerce any of them into paying for a party campaign. Since the party appeared as if it could now be of little assistance to their campaigns, they opted out of the original arrangements, even though the new technologies were of little use to them at the time. By the late 1960s only Byron Rogers's financial contributions in partisan elections and the *de facto* partisan style of mayoral contests were sustaining the party.

Sudden collapses in the supply of campaign resources, however, such as occurred after the abandonment of the multi-member districts in Denver, were rather unusual. The more gradual causes of resource decline – demographic changes, the decline of the labour unions, and shifts in activist recruitment – were far more widespread. This suggests that the conservative response by candidates might have been far more common and that, without the technological revolution, there might have been many more instances of party-centred campaigning surviving into the late 1970s. For example, without the new technologies, it would have been far more difficult for mayoral candidates in New York

City to have separated themselves from the ties of party, and their separation was instrumental in weakening party resources throughout the city's Democratic parties. In these circumstances no mayoral candidate could have been as independent of his party as Edward Koch was in 1977; the limits of independence would have been much more like those facing Robert Wagner in 1961. The constraints would have been even greater for candidates for lesser offices. This suggests that the era of organized factional slates, last seen at the city level in New York's Democratic primaries in the mid-to-late 1960s might have survived much longer. Equally, at the level of state politics, we would probably have seen that, in those states where the party organization could still influence the nomination, it would have been far more difficult for insurgents to win nominations. With the technology of 1946, insurgents such as the text-book example of Milton Shapp in Pennsylvania in 1966 would have been a much less serious threat to Democratic party leaders there. To conclude: there do not seem to be strong grounds for arguing that the technological changes were a necessary condition for party decline. While some parties might have declined much more slowly but for these changes, it seems implausible to claim that this would have happened to all of them, or that, in the very long term even in places such as New York, party leaders could have retained the role they played in the 1940s.

But what of the upheavals of the 1960s? Either directly or indirectly, how much responsibility can be attached to the issue conflict and issue 'extremism' of that decade in promoting party collapse? It is undeniable that in some cases, including that of the East Bay which we have discussed at length, the transformation of the party's role is related to the '60s issues'. But what collapsed there were extra-legal party structures operating where there were no legally defined parties below the level of the county. While a complete account of party collapse in America must embrace these developments, it would be highly misleading to claim that their impact was as devastating under other conditions. Nevertheless, there are two important respects in which the issue conflicts did harm the Democratic parties. First, they helped to make issue-oriented activists much more sceptical about the value of party; what emerged in the 1960s was issue-activism which was not party-oriented, as it was in the 1950s, but which was prepared to use party institutions for realizing objectives as, and when, they seemed useful. Attachments to individual candidates, and the use of initiative referendums and other devices, weakened the role of the Democratic party as an organizer of activist groups. Secondly, the issue conflicts actually revived long-standing anti-party sentiments in America, sentiments which were minority ones in the amateur Democrat movement of the 1950s, but which became more apparent in the late 1960s. Whether by accident, or by the design of individuals who supported the break-up of the parties, institutional reforms were introduced in the turmoil of the 1960s which weakened the power of parties in the nomination process. This happened in New York in 1967, but the most important reforms were those at the federal level concerning presidential selection in the Democratic party. The result of these reforms has been called the 'plebiscitary nomination system' by James Ceaser.[2]

Two points about the relevance of this consequence of the issue conflicts of the 1960s seem uncontentious. The presidential nomination process did change radically, and the vast reduction in the role party leaders played in this process probably did weaken further the parties at the local level. But compared with the changes in campaign technology, and the other social, economic, and political changes which we have mentioned, the effects of these reforms on local and state parties were secondary. They were enacted after these other forces had severely weakened the parties. Indeed, it could be argued, though this is not the place to do so, that some form of decentralized nomination system was bound to emerge in the 1970s and 1980s, given the transformation which the parties had otherwise experienced. In a world without party bosses, the 'mixed' presidential nominating system of the mid-twentieth century would have been modified to fit the changed circumstances. Yet, as a catalyst in prompting a more rapid transformation of Democratic parties, the issue conflicts of the 1960s must not be underestimated. A man with serious heart and bronchial ailments is not going to recover more easily if he is involved in a major road accident. So, to speak figuratively, it was with the Democratic parties. Both the technological and the 'long-term' changes we have discussed would have transformed the parties eventually. Interacting with each other, they produced an extraordinarily rapid transformation, which the particular circumstances of the 1960s made yet more rapid. Thus it was that, in a period of little more than ten years, America's Democratic parties came to resemble no longer the institutions described in textbooks in 1962 or 1963. This transformation was to become more complete during the rest of the 1970s. By the end of that decade party organizations were far weaker than they had ever been since the emergence of mass democracy in the 1820s; party identification in the electorate was so weak that political scientists had invented the new concept of electoral dealignment. In one sense, at least, the party had died.

If it is accepted, however, that a major transformation has occurred, there remains the question of establishing *what* has happened to America's parties. If we describe the parties as having declined or collapsed, does this not mean, it might be asked, that there is no longer party politics in America? If so, how are we to reconcile this with the fact that the terms 'Democrat' and 'Republican' are seen as much in newspaper reports of politics in the 1980s as they were in the early 1960s? In considering this issue, we are brought back to the brief discussion of the concept of party we introduced in Chapter 1. Perhaps the most useful way of proceeding is to explain how genuinely party-less politics could have developed from caucus-cadre parties.

For a set of elected politicians to be regarded as a party, there have to be ties which bind the representatives to each other and to the electoral activists who helped to elect them. Now we must not commit the common European fallacy and assume that in a modern state these ties must be ones primarily of shared policy views or ideology. They may not be. We would still identify a 'party' as a party if the main ties were ones of personal loyalty (as, say, in Japan) or if they arose from a need to share scarce electoral resources. America's parties began like the oldest European parties, as collections of local notables who would get together to help elect representatives. Their relations with other groups of notables, and the relations between the party's representatives, were always loose. In America the caucus-cadre type of party survived the emergence of mass electorates, arguably because its decentralized structure was well suited to the peculiarly decentralized political system, while the more modern branch type was not. Nevertheless, the caucus-cadre parties did not survive in their original form, even in America, where there were three major developments in their structure between the mid-nineteenth century and the early twentieth century. First, Irish entrepreneurs in many cities devised an organizational form which provided the parties with the large work-forces they needed to mobilize mass electorates. Secondly, state governments then regulated the organizational form of the parties, providing them with a fixed, formal structure from the lowest level of politics to the state level. Finally, these same governments introduced some form of primary election, as the means by which major parties were to select their candidates. Eventually, as we have seen, the last of these developments made it possible for candidates to appeal above the heads of the party leaders to the primary electorate when seeking nomination; changes in campaign technology made this relatively easy for major office candidates, for insurgents no longer had to acquire an army of campaign workers of their own.

To the extent that they depended neither on local party notables nor on their fellow office-holders for gaining nomination and election, candidates (the new notability) could then be in a position to practise party-less politics. While they would have to adopt a party label for the purpose of winning elections, their electoral independence could put them in a position where, both in office and in campaigning, they would rely no more on co-operation with their fellow wearers of the party label than they would on help from anyone who took the other party's label. This would be party-less politics because, although there would be party labels in regular use, there would not be any ties to bind the politicians together as a party. Thus, the introduction of the primary election into the American caucus-cadre parties did provide the potential for the disappearance of party politics. In fact, this has not happened, nor does it appear that this outcome is likely – at least in most of the non-southern states.

There are three reasons why the collapse of the Democratic parties has not generated a transformation into genuinely party-less politics. In the first place, although virtually all of the highly centralized party organizations of the pre-war period have disappeared, some aspects of this older style have endured. For example, there are ties which bind most state legislators in New York City to at least some part of the formal party structures. These ties are much weaker than they were but they have not been eliminated completely. Similarly, even with Edward Koch as mayor, City Hall did not ignore all the borough parties: some association with them was necessary in the interest of effective management. Secondly, as we saw in Chapter 6, under some conditions electoral co-operation between candidates could develop even in states with a tradition of individual campaigning by candidates. The keys to this development seem to have been (i) institutional arrangements which centralized power in the state legislature and (ii) the high costs of campaigning at some elections. Of course, the conditions favourable to the growth of 'warlordism' in California were highly unusual, and no one could possibly claim that this will be a model for relations between candidates elsewhere. Nevertheless, it does expose the problems facing candidates for lesser offices in an era when campaign resources are expensive. Even when only a few election campaigns in his career are extremely expensive, the candidate may have to accept a little help from potential party friends; out of such necessities some party ties are forged. Finally, and most important of all, there is a consideration which we have not yet discussed – the role of the Republican parties. To understand the effect that Republican parties might have on the development of party-less politics, it is useful to consider, by way of analogy, an argument which the economist P. W. S. Andrews called the 'Ward–Perkins point'.

In his discussion of the requirement for universal rationality among consumers in economic theory, Andrews argued:

It seems that one very simple consideration will suffice to establish how far from necessity it is to require a universal rationality of consumer choice. A minority of rational buyers at any time would suffice to make it profitable for the urban retailer, in the matter of the quality of his goods and the prices which he puts on them, not to behave very differently from how he would behave *if* consumers generally were rational. I myself call this the 'Ward–Perkins point' . . .[3]

Economic competition will produce similar results when only a few consumers are rational and when all of them are; the rationality of the few imposes order on the market. Now a parallel point may be made about competitive political parties. If one party in a two-party system is not experiencing the kinds of pressures which produce genuinely party-less politics, then it will limit the collapse of the other party into merely a label, irrespective of the strength of the factors promoting disintegration in that party. In other words, if the Republican party was to maintain the campaign links between candidates and party organizations, and retain its more conservative issue stances, it would restrict the potential for the Democratic party becoming a completely 'catch-all' party without any ties of loyalty or issue coherence. If the Republican party is not being reduced to a label, neither can the Democratic party be reduced to one. Disproportionately, liberal activists and potential candidates would gravitate towards the Democrats, and this would give the party some kind of identity with respect to public policies. There would also be barriers to public officials changing their party labels, because the convert to Republicanism would have to be certain that he would have enough in common with Republican activists in his area. In short, one party-like 'party' will prevent the other party from becoming a Tweedledee to its Tweedledum.

In fact, there is evidence that Republican parties have played this role in retaining order in the party systems of many American states. Nexon's study of the period 1956–64 – the years immediately preceding the transformation of the Democratic parties – showed that Republican activists were far more likely to be oriented to conservative issue positions than their Democratic counterparts, and unlike the Democrats they were more extreme than their own party's voters.[4] Moreover, Nexon reported that activist participation was much higher than in the Democratic parties – a factor which would make them a more potent force in the nomination process. In the 1960s and 1970s the grip of conservative Republican activists tightened: they ousted a number of moderate officials from their county parties, so that the liberal, largely north-eastern, wing of the party declined in influence. Only in US Senate elections, where it was easiest for candidates to mobilize independently of party, has this kind of Republicanism retained the level of representation in elected public office which it had earlier. Because Republican parties have been much more effective as fund-raisers than Democratic parties, and have distributed this money to suitable candidates, the party link has been preserved. The result is that, even in weak party states such as California, Republican representatives in the state legislature act as a party and most of them are noticeably more conservative in their views than the Democratic legislators. Competition between the parties at the state level, therefore, seems to have prevented the collapse of many Democratic parties into mere labels.

In pointing out the crucial role which the Republican parties appear to have played, inevitably we raise a question which falls largely outside our study, Why should the two parties have been so very different, when superficially their structures are so similar? There are at least two explanations of this apparent paradox. The first points to the logic of the position of a minority party in a political system which tends to produce electoral realignments in which there is usually a 'sun' and a 'moon' party.[5] Ghettoized in its minority position, the 'moon' party will tend to consolidate its position, rather than seek to expand its activist base and broaden its issue appeal, because the latter is a strategy which has little hope of success. In this view, the Republican party became an ideological party because it lost out in the realignment of the early 1930s, and it retained its character when electoral dealignment replaced the earlier periodic realignments. The plausibility of this explanation depends on an adequate account being given of why the 'sun' parties do not shrink towards the size of a minimum winning coalition.[6] The second explanation relates the American party system to its socio-economic structure. The dominance of business groups, and of the capitalist ethic, meant that an industrialized America could support a major party which drew primarily on this narrow range of interests. There was no such single powerful set of interests which the Democrats could mobilize against it; consequently, the Democratic party had to envelop many more diverse groups and become much more of a coalition of interests than the Republican party. This account might also draw attention to the role of the plurality voting system in preserving this coalition party from fragmentation into a number of smaller parties. These brief sketches of possible lines of argument draw attention to an obvious point – the complexity of the issues involved in providing an adequate account of the paradox.

Of course, to claim that the Republican parties are partly responsible for the non-emergence of genuinely party-less politics in America is not to overlook two obvious points. First, Republicans have embraced the new campaign technologies, and in many cases have used them much more extensively than Democrats; Republican parties are also far from being European branch-type parties. Nevertheless, the informal links between Republican notables and candidates, together with a more uniformly ideological primary electorate, do seem to have served as effective constraints on the ideology and non-party orientations of their candidates. Secondly, the transformation which the Democratic parties have undergone is significant, even though a full metamorphosis to party-less politics may seem unlikely. Governing at all levels of politics in America has become more difficult for Democrats, because of the collapse of the ties between political élites which the parties previously provided. The complete failure of President Carter to lead his party in Congress, when on paper he had a large majority of supporters there, is merely one manifestation of the changes which have occurred. Similar tensions between Democratic executives and legislatures can be found at the state level.

In suggesting that we must examine the role of the Republican parties when trying to explain why party-less politics did not develop, we come finally to an issue raised in the first chapter – the a-theoretical nature of most studies of American parties. If political scientists are to provide an adequate account of the persistence of parties under seemingly adverse conditions, then they must address the problem of the nature of American parties and the American party systems. We need to understand how parties, different from those usually found in western societies, can survive. It is only with an understanding of the interaction of social cleavages, governmental structures, and electoral laws that we can hope to explain the dynamics of party systems, and thereby explain the major, if incomplete, transformation of the Democratic parties. In this connection, there has been some important research conducted – for example, by Walter Dean Burnham.[7]

As we suggested in Chapter 1, however, the study of the parties themselves has been dominated by an a-theoretical approach which has emphasized the collection of quantifiable data on a limited range of matters, at the expense of macro-political theory. If the isolation of American political science has been an important element in this lack of research on many central issues, European political science must also share some of the blame. Widespread European contempt for the non-ideological American parties has not helped to stimulate debate in the comparative study of parties – rather, American, and Japanese, parties have simply been treated as curiosities. Moreover, there has been a general lack of interest among European scholars in the American political system. In Britain there are, perhaps, twenty-five or so full-time academics who are engaged actively in research on American politics, whilst in the rest of Europe there seems to be even less expertise.[8] The tradition of de Tocqueville, Bryce, and Brogan has largely died, and recent European contributions to the study of American politics have been relatively small. The fact is, that if European scholars were to concern themselves with America, they might be able to make significant contributions to the understanding of a political system which is very different from any of their own.

## Notes

[1] See Chapter 1, note 24.

[2] Ceaser, *Presidential Selection*.

[3] P. W. S. Andrews, *On Competition in Economic Theory* (London: Macmillan, 1964), 101–2.

[4] Nexon, 'Asymetry in the Political System: Occasional Activists in the Republican and Democratic Parties, 1956–1964'.

[5] The analogy was originally introduced by Samuel Lubell, *The Future of American Politics*, rev. edn. (Garden City, NY: Doubleday Anchor Books, 1956), p. 212.

[6] The argument that a political coalition will tend to decline in size, to that of a minimum winning coalition, is developed in William H. Riker, *The Theory of*

*Political Coalitions* (New Haven: Yale University Press, 1962). Riker cites examples of 'over-sized' party coalitions in American political history in demonstrating the 'size principle'. However, while it is clear that his examples are ones of large coalitions declining in size at subsequent elections, it is much less obvious that in all the cases the decline produced two parties with coalitions of a similar size.

[7] For a complete list of his publications see Burnham, *The Current Crisis in American Politics*.

[8] For example, at two recent conferences in continental Europe at which I gave papers on American politics, I encountered only two non-British participants (out of a total of about fifty) who had anything more than a very basic knowledge of American government.

# Bibliography and Other Sources

## Manuscripts

Minutes and Records of the Denver Democratic Party

## Books

Adler, Norman M., and Blank, Blanche D., *Political Clubs in New York* (New York: Praeger, 1975)

Agranoff, Robert (ed.), *The New Style in Election Campaigns*, 2nd edn. (Boston: Holbrook Press, 1976)

Alexander, Herbert E., *Financing Politics*, 2nd edn. (Washington. DC: Congressional Quarterly Press, 1980)

Andrews, P. W. S., *On Competition in Economic Theory* (London: Macmillan, 1964)

Bain, George Sayers, and Price, Robert, *Profiles of Union Growth* (Oxford: Basil Blackwell, 1980)

Barone, Michael, and Ujifusa, Grant, *The Almanac of American Politics 1982* (Washington, DC: Barone, 1982)

Bibby, John F., Mann, Thomas E., and Ornstein, Norman J., *Vital Statistics on Congress 1980* (Washington, DC: American Enterprise Institute, 1980)

Bok, Derek C., and Dunlop, John T., *Labor and the American Community* (New York: Simon and Schuster, 1970)

Broder, David, *The Party's Over* (New York: Harper and Row, 1972)

Burnham, Walter Dean, *Critical Elections and the Mainsprings of American Politics* (New York: Norton, 1970)

——, *The Current Crisis in American Politics* (New York: Oxford University Press, 1982)

Campbell, Angus, Converse, Philip, Miller, Warren, and Stokes, Donald, *The American Voter* (New York: Wiley, 1960)

——, ——, ——, and ——, *Elections and the Political Order* (New York: Wiley, 1960)

Carney, Francis, *The Rise of the Democratic Clubs in California* (New York: Holt, 1958)

Ceaser, James, *Presidential Selection* (Princeton: Princeton University Press, 1979)

Congressional Quarterly, *C. Q. Guide: Current American Government, Spring 1980* (Washington, DC: Congressional Quarterly Press, 1979)
Costikyan, Edward N., *Behind Closed Doors* (New York: Harcourt, Brace, 1966)
——, *How to Win Votes* (New York: Harcourt, Brace, 1980)
Crotty, William J., *Decision for the Democrats* (Baltimore: Johns Hopkins Press, 1978)
Dexter, Lewis A., *The Sociology and Politics of Congress* (Chicago: Rand McNally, 1969)
Duverger, Maurice, *Political Parties*, 2nd English edn. (London: Methuen, 1959)
Elazar, Daniel J., *American Federalism*, 2nd edn. (New York: Crowell, 1972)
Eldersveld, Samuel J., *Political Parties: A Behavioral Analysis* (Chicago: Rand McNally, 1968)
Epstein, Leon D., *Political Parties in Western Democracies* (New York: Praeger, 1968)
Fenno, Richard F., *Home Style* (Boston: Little, Brown, 1978)
Fishel, Jeff (ed.), *Parties and Elections in an Anti-Party Age* (Bloomington, Ind., Indiana University Press, 1978)
Fox, Harrison W., Jr., and Hammond, Susan Webb, *Congressional Staffs* (New York: Free Press, 1977)
Graber, Doris A., *Mass Media and American Politics* (Washington, DC: Congressional Quarterly Press, 1980)
Guterbock, Thomas S., *Machine Politics in Transition* (Chicago: University of Chicago Press, 1980)
Hamilton, Charles V., *The Black Preacher in America* (New York: Morrow, 1978)
Harrigan, John J., *Political Change in the Metropolis* (Boston: Little, Brown, 1976)
Hershey, Marjorie Randon, *The Making of Campaign Strategy* (Lexington, Mass.: D. C. Heath, 1974)
Hirsch, Fred, *Social Limits to Growth* (London: Routledge and Kegan Paul, 1977)
Hirschman, Albert O., *Shifting Involvements* (Oxford: Martin Robertson, 1982)
Jacobson, Gary C., and Kernell, Samuel, *Strategy and Choice in Congressional Elections* (New Haven and London: Yale University Press, 1983)
Jewell, Malcolm E., *Representation in State Legislatures* (Lexington: University of Kentucky Press, 1982)
——, and Olson, David M., *American State Political Parties and Elections* (Homewood, Ill.: Dorsey Press, 1978)

Katznelson, Ira, *City Trenches* (New York: Pantheon, 1981)
Keech, William R., and Matthews, Donald R., *The Party's Choice* (Washington, DC: Brookings, 1976)
Kelley, Stanley, *Professional Public Relations and Political Power* (Baltimore: Johns Hopkins Press, 1956)
Kessel, John H., *The Goldwater Coalition* (Indianapolis: Bobbs-Merrill, 1968)
Key, V. O., *American State Politics* (New York: Knopf, 1956)
——, *Politics, Parties and Pressure Groups*, 4th edn. (New York: Crowell, 1958)
King, Anthony, *The New American Political System* (Washington, DC: American Enterprise Institute, 1978)
Kuttner, Robert, *Revolt of the Haves* (New York: Simon and Schuster, 1980)
Lee, Eugene C., *The Politics of Nonpartisanship* (Berkeley: University of California Press, 1960)
Leuthold, David, *Electioneering in a Democracy* (New York: Wiley, 1968)
Lowi, Theodore J., *At the Pleasure of the Mayor* (New York: Free Press, 1964)
Lubell, Samuel, *The Future of American Politics* rev. edn. (Garden City, NY: Doubleday Anchor Books, 1956)
Mackintosh, John P., *The British Cabinet* (London: Stevens, 1962)
Malbin, Michael T. (ed.), *Parties, Interest Groups and Campaign Finance Laws* (Washington, DC: American Enterprise Institute, 1980)
Mayhew, David R., *Congress: The Electoral Connection* (New Haven and London: Yale University Press, 1974)
Morris, Charles R., *The Cost of Good Intentions* (New York: Norton, 1980)
Nelsen, Hart M., and Nelsen, Anne Kusener, *Black Churches in the Sixties* (Lexington: University of Kentucky Press, 1975)
Nie, Norman H., Verba, Sidney, and Petrocik, John R., *The Changing American Voter* (Cambridge, Mass.: Harvard University Press, 1976)
Nimmo, Dan, *The Political Persuaders* (Englewood Cliffs, NJ: Prentice-Hall, 1970)
Novak, Michael, *The Rise of the Unmeltable Ethnics* (New York: Macmillan, 1971)
Ogg, Frederick A., and Ray, P. Norman, *Introduction to American Government*, 7th edn. (New York: Appleton-Century, 1942)
Peel, Roy V., *The Political Clubs of New York City* (New York: Putnam, 1935)
Polsby, Nelson W., *Consequences of Party Reform* (Oxford and New York: Oxford University Press, 1983)

Pomper, Gerald M. (ed.), *Party Renewal in America* (New York: Praeger, 1980)

Ranney, Austin, *Curing the Mischiefs of Faction* (Berkeley: University of California Press, 1975)

Riker, William H., *The Theory of Political Coalitions* (New Haven: Yale University Press, 1962)

Sabato, Larry J., *The Rise of Political Consultants* (New York: Basic Books, 1981)

Sartori, Giovanni, *Parties and Party Systems* (Cambridge: Cambridge University Press, 1976)

Sayre, Wallace S., and Kaufman, Herbert, *Governing New York City* (New York: Russell Sage Foundation, 1960)

Scarrow, Howard A., *Parties, Elections and Representation in the State of New York* (New York and London: New York University Press, 1983)

Schattschneider, E. E., *The Semisovereign People* (Hinsdale, Ill.: Dryden Press, 1975)

Schlesinger, Joseph A., *Ambition and Politics* (Chicago: Rand McNally, 1966)

Scott, Ruth K., and Hrebenar, R. J., *Parties in Crisis* (New York: Wiley, 1979)

Shafer, Byron E., *Quiet Revolution* (New York: Russell Sage Foundation, 1983)

Sorauf, *Party Politics in America*, 4th edn. (Boston: Little, Brown, 1980)

Stanley, David T., *Managing Local Government Under Union Pressure* (Washington, DC: Brookings, 1972)

Stave, Bruce M., *The New Deal and the Last Hurrah* (Pittsburgh: University of Pittsburgh Press, 1970)

Tolchin, Martin, and Tolchin, Susan, *To the Victor* (New York: Random House, 1971)

Turner, Julius, *Party and Constituency* (Baltimore: Johns Hopkins Press, 1951)

Verba, Sidney, and Nie, Norman H., *Participation in America* (New York: Harper and Row, 1972)

Ware, Alan, *The Logic of Party Democracy* (London: Macmillan, 1979)

Wattenberg, Martin P., *The Decline of American Political Parties, 1952–1980* (Cambridge, Mass. and London: Harvard University Press, 1984)

Wilson, Graham K., *Unions in American National Politics* (London: Macmillan, 1979)

Wilson, James Q., *The Amateur Democrat* (Chicago: University of Chicago Press, 1962)

——, *Political Organizations*, (New York: Basic Books, 1973)

Young, William H., *Ogg and Ray's Introduction to American Government*, 12th edn. (New York: Appleton-Century, 1962)

**Articles**

Abramowitz, Alan, McGlennon, John, and Rapoport, Ronald, 'The Party Isn't Over: Incentives for Activism in the 1980 Presidential Nominating Campaign', *Journal of Politics*, 45 (1983), 1006–15

Adkisson, John, 'Staff Assistant Today, Assembly Member Tomorrow', in T. Anthony Quinn and Ed Salzman (eds.), *California Public Administration* (Sacramento: California Journal Press, 1978)

Adrian, Charles R., 'A Typology for Nonpartisan Elections', *Western Political Quarterly*, 12 (1959), 449–58

Axelrod, Robert, 'Where the Votes Come From: An Analysis of Electoral Coalitions, 1952–1968', *American Political Science Review*, 66 (1972), 11–20

Burnham, Walter Dean, 'The End of American Party Politics', *Trans Action*, 7 (1969), 12–22

Clark, Peter B., and Wilson, James Q., 'Incentive Systems: A Theory of Organization', *Administrative Science Quarterly*, 6 (1961), 129–66

Committee on Political Parties of the American Political Science Association, 'Toward a More Responsible Two-Party System', *American Political Science Review*, 44 (1950), Supplement.

Cotter, Cornelius P., and Bibby, John F., 'Institutional Development of Parties and the Thesis of Party Decline', *Political Science Quarterly*, 95 (1980), 1–28

Dennis, Jack, 'Trends in Public Support for the American Party System', *British Journal of Political Science*, 5 (1975), 187–230

Fiorina, Morris P., 'The Case of the Vanishing Marginals: The Bureaucracy Did It', *American Political Science Review*, 71 (1977), 177–81

——, 'The Decline of Collective Responsibility in American Politics', *Daedalus*, 109 (1980), 25–45

Gibson, James L., Cotter, Cornelius P., Bibby, John F., and Huckshorn, Robert J., 'Assessing Party Organizational Strength', *American Journal of Political Science*, 27 (1983), 193–222

Gifford, Bernard R., 'New York City and Cosmopolitan Liberalism', *Political Science Quarterly*, 93 (1978–9), 559–84

Goodman, T. William, 'How Much Political Party Centralization do We Want?', *Journal of Politics*, 13 (1951), 536–61

Greenstone, J. David, 'Party Pressure on Organized Labor in Three Cities', in M. Kent Jennings and L. Harmon Zeigler (eds.), *The Electoral Process* (Englewood Cliffs, NH: Prentice-Hall, 1966)

Grillo, Evelio, 'D. G. Gibson: A Black Who Led the People and Built the

Democratic Party in the East Bay', in Harriet Nathan and Stanley Scott (eds.), *Experiment and Change in Berkeley* (Berkeley: Institute of Governmental Studies, University of California, 1978)

Hamilton, Charles V., 'The Patron-Recipient Relationship and Minority Politics in New York', *Political Science Quarterly*, 94 (1979), 211–27

Hansmann, Henry B., 'The Role of Non-profit Enterprise', *Yale Law Journal*, 89 (1980), 835–901

Hofstetter, C. Richard, 'The Amateur Politician: A Problem in Construct Validation', *Midwest Journal of Political Science*, 15 (1971), 31–56

Horton, Raymond D., 'Municipal Labor Relations in New York City', in Robert H. Connery and William V. Farr, *Unionization of Municipal Employees* (New York: The Academy of Political and Social Science, 1970)

Huckshorn, Robert J., and Bibby, John F., 'State Politics in an Era of Political Change', in the American Assembly (ed.), *The Future of American Political Parties* (Englewood Cliffs, NJ: Prentice-Hall, 1982)

Kayden, Xandra, 'The Nationalizing of the Party System', in Michael J. Malbin (ed.), *Parties, Interest Groups and Campaign Finance Laws*, (Washington, DC: American Enterprise Institute, 1980)

Kent, T. J., Jr., 'Berkeley's First Liberal Democratic Regime, 1961–1970: The Postwar Awakening of Berkeley's Liberal Conscience', in Harriet Nathan and Stanley Scott (eds.), *Experiment and Change in Berkeley* (Berkeley: Institute of Governmental Studies, University of California, 1978)

King, Anthony, 'Political Parties in Western Democracies', *Polity*, 2 (1969), 111–41

Klain, Maurice, 'A New Look at the Constituencies', *American Political Science Review*, 49 (1955), 1005–19

Lockard, W. Duane, 'Legislative Politics in Connecticut', *American Political Science Review*, 48 (1954), 166–73

Loughlin, Walter P., 'Election Administration in New York City: Pruning the Political Thicket', *Yale Law Journal*, 84 (1974), 61–85

Malbin, Michael, 'Of Mountains and Molehills: PACs, Campaigns and Public Policy', in Michael Malbin (ed.), *Parties, Interest Groups and Campaign Finance Laws* (Washington, DC: American Enterprise Institute, 1980)

Nexon, David, 'Asymetry in the Political System: Occasional Activists in the Republican and Democratic Parties', *American Political Science Review*, 65 (1971), 716–30

Parenti, Michael, 'Ethnic Politics and the Persistence of Ethnic Identification', *American Political Science Review*, 61 (1967), 717–26

Polsby, Nelson W., 'Mayor Daley and the Bishop', *Political Studies*, 28 (1980), p. 465

Pressman, Jeffrey L., 'Preconditions of Mayoral Leadership', *American Political Science Review*, 66 (1972), 511–24

——, and Sullivan, Dennis G., 'Convention Reform and Conventional Wisdom: An Empirical Assessment of Democratic Party Reforms', *Political Science Quarterly*, 89 (1974), 539–62

Protess, David L., and Gitelson, Alan R., 'Political Stability, Reform Clubs and the Amateur Democrat', in William J. Crotty (ed.), *The Party Symbol* (San Francisco: W. H. Freeman, 1980)

Ranney, Austin, 'Towards a More Responsible Two-Party System: A Commentary', *American Political Science Review*, 45 (1951), 488–99

——, 'Parties in State Politics', in Herbert Jacob and Kenneth N. Vines (eds.), *Politics in the American States*, 1st edn. (Boston: Little, Brown, 1965)

——, 'Parties in State Politics', in Herbert Jacob and Kenneth N. Vines (eds.), *Politics in the American States*, 3rd edn. (Boston: Little, Brown, 1976)

——, 'The Political Parties: Reform and Decline', in Anthony King (ed.), *The New American Political System* (Washington, DC: American Enterprise Institute, 1978)

Salisbury, Robert H., 'The Urban Party Organization Member', *Public Opinion Quarterly*, 29 (1965–6), 550–64

Shefter, Martin, 'Regional Receptivity to Reform: The Legacy of the Progressive Era', *Political Science Quarterly*, 98 (1983), 459–83

Stokes, Donald E., 'Parties and the Nationalization of Electoral Forces', in William Nisbet Chambers and Walter Dean Burnham (eds), *The American Party Systems* (New York: Oxford University Press, 1967)

Tolchin, Susan, and Tolchin, Martin, 'How Judpeships Get Bought', *New York Magazine*, 15 Mar. 1971

Ware, Alan, 'The End of Party Politics? Activist-Officeseeker Relationships in the Colorado Democratic Party', *British Journal of Political Science*, 9 (1979), 237–50

——, 'Why Amateur Party Politics has Withered Away: The Club Movement, Party Reform and the Decline of American Party Organizations', *European Journal of Political Research*, 9 (1981), 219–36

——, 'Party Decline and Party Reform', *Teaching Politics*, 12 (1983), 82–96

Wildavsky, Aaron B., 'The Goldwater Phenonemon: Purists, Politicians and the Two-Party System', *Review of Politics*, 27 (1965), 386–413

Wilson, James Q., 'Negro Politics in the North', in Harry A. Bailey, Jr. (ed.), *Negro Politics in America* (Columbus: Merrill, 1967)

Wolfinger, Raymond E., 'The Development and Persistence of Ethnic

Voting', *American Political Science Review*, 59 (1965), 896–908
——, 'Why Political Machines Have Not Withered Away and Other Revisionist Thoughts', *Journal of Politics*, 34 (1972), 365–98

## Unpublished Papers

Cotter, Cornelius P., Gibson, James L., Bibby, John F., and Huckshorn, Robert J., 'State Party Organizations and the Thesis of Party Decline', paper presented at the Annual Meeting of the American Political Science Association, Washington, DC, 1980
Goetchus, V., 'The Village Independent Democrats', unpublished manuscript, 1963
Pinto-Duschinsky, Michael, 'Theories of Corruption in American Politics', paper presented at the Annual Meeting of the American Political Science Association, Chicago, 1976
Weinberg, Lee S., Margolis, Michael, and Ranck, David F., 'Local Party Organization: From Disaggregation to Disintegration', paper presented at the Annual Meeting of the American Political Science Association, Washington, DC, 1980

## Newspapers and Periodicals

*Amsterdam News*
*California Journal*
*Colorado Democrat*
*Congressional Quarterly Weekly Reports*
*Denver Post*
*New York Magazine*
*New York Times*
*Montclarion*
*Oakland Tribune*
*Rocky Mountain News*
*San Francisco Chronicle*
*Village Voice*

## Interviews

One hundred and thirty-five interviews were conducted with politicians, political activists, and others between October 1978 and August 1982. All interviewees were told that, if they wished, no information or opinions they presented to me would be made public in a form which would lead to them being identified. A sufficient number of the interviewees asked for this condition to be respected for it to be decided that all of them should be identified by a number, and not by name, in any publication. This is how they are presented here. With seven exceptions, all the interviews were taped on cassettes; the interviews usually lasted between thirty and ninety minutes. These cassettes are stored at the University of Warwick. Use was also made of some of the forty-eight taped interviews which were conducted in Denver in 1976 on an earlier research project. These tapes are also stored at the University.

## Surveys

Two surveys, using questionnaires sent by mail, were conducted during the research.

The first survey was financed by the Social Science Research Council, and the data from this are available from the ESRC's Data Archive. In 1979 questionnaires were sent to all 100 members of the Denver Democratic Party's Executive Committee, and a similar questionnaire was sent to 400 precinct committee members. (There are approximately 1,000 precinct committee members in the city party.) The first mailing was sent in April 1979, and a follow-up mailing was sent a month later. The response rate was 52 per cent among the Executive Committee members and 51 per cent among precinct committee members.

The second survey involved the sending of questionnaires to the ninety-five members from New York City serving in the state Assembly and Senate in March 1982. The first mailing was sent in May 1982 and a follow-up mailing was sent a month later. Sixty-six legislators responded to this questionnaire. The completed questionnaires are stored at the University of Warwick.

# INDEX